Mathematical Computation with Maple V: Ideas and Applications

Mathematical Computation with Maple V: Ideas and Applications

Proceedings of the Maple Summer Workshop and Symposium, University of Michigan, Ann Arbor, June 28–30, 1993

Thomas Lee

Editor

BIRKHÄUSER

Boston • Basel • Berlin

Thomas Lee
Waterloo Maple Software
Waterloo, Ontario
CANADA N2L 5J2

Printed on acid-free paper

© Birkhäuser Boston 1993   *Birkhäuser*

Maple and Maple V are registered trademarks of Waterloo Maple Software.

ISBN 0-8176-3724-9
ISBN 3-7643-3724-9

Camera-ready text prepared by the Authors.
Printed and bound by Grafacon, Hudson, MA
Printed in the U.S.A.

9 8 7 6 5 4 3 2 1

Contents

PREFACE

Developments in both computer hardware and software over the decades have fundamentally changed the way people solve problems. Technical professionals have greatly benefited from new tools and techniques that have allowed them to be more efficient, accurate, and creative in their work.

Maple V and the new generation of mathematical computation systems have the potential of having the same kind of revolutionary impact as high-level general purpose programming languages (e.g. FORTRAN, BASIC, C), application software (e.g. spreadsheets, Computer Aided Design - CAD), and even calculators have had. Maple V has amplified our mathematical abilities: we can solve more problems more accurately, and more often. In specific disciplines, this amplification has taken excitingly different forms.

For mathematicians, computer algebra systems has spawned entire new research and application directions. In addition, Maple V has allowed mathematicians to approach existing problems in ways that were not conceivable 20 years ago.

For scientists and engineers, Maple V represents a bridge between theory and practice. Techniques that were traditionally considered too complex for application are now beginning to be used by a wide range of scientists and engineers in both academia and industry. Indeed the concept of "scientific computation" has evolved from a relatively narrow field encompassing primarily numerical methods to a much broader field that includes, numerical methods, algebraic and analytical methods, and graphics.

Perhaps the greatest impact has been felt by the education community. Today, it is nearly impossible to find a college or university that has not introduced mathematical computation in some form, into the curriculum. Students now have regular access to the amount of computational power that were available to a very exclusive set of researchers five years ago. This has produced tremendous pedagogical challenges and opportunities.

Comparisons to the calculator revolution of the 70's are inescapable. Calculators have extended the average person's ability to solve common problems more efficiently, and arguably, in better ways. Today, one needs at least a calculator to deal with standard problems in life - budgets, mortgages, gas mileage, etc. For business people or professionals, the spreadsheet is quickly becoming the minimum level of electronic assistance for solving modern problems. Many look towards the next decade with great anticipation as what is now considered advanced mathematical and computing techniques (like Maple V) begin influencing the work of businessmen, social scientists, and others who have traditionally not exploited the tremendous potential of this particular dimension of the Information Revolution.

The Maple Summer Workshop and Symposium - MSWS '93

The primary goal of MSWS '93 is to bring interested people together and to generate ideas and strategies to promote the effective use of Maple V and other modern computation techniques. This year, the conference was held on June 28-30 at the University of Michigan in

Ann Arbor. This volume summarizes the paper sessions of the conference. The paper presentations were part of a varied conference program designed to encourage interaction among users from a wide variety of disciplines.

One of the most encouraging aspects of the paper sessions was the impressive breadth. It is a clear demonstration of the influence that Maple V is having on the research and academic community. As important as the specific technical topics are, it is equally important to note the differences in basic approaches to Maple V-based problem solving. The collection includes a wide range of perspectives - from "computerization" of classical theory to facilitate more efficient and accurate problem-solving, to the development of complete Maple V-based systems for the automatic solution of problems in ways that are not possible with conventional computing systems. If this volume is any indication of the future of mathematical computing, then it is very clear that we will witness an even more impressive proliferation of computer algebra in the years to come.

MSWS '93 and this volume could not have been achieved without the invaluable assistance of many dedicated and enthusiastic people. The Editor would like to express his gratitude to the reviewers who served to ensure overall quality in the papers. Furthermore, the assistance of Paola D'Alessandro, Karin Turner, Jeff Watling, Lee Liming, Melanie McInness, Ann Kostant and Edwin Beschler is greatly appreciated.

T. Lee
Waterloo, Canada

I

MAPLE V IN EDUCATION

INTRODUCTORY QUANTUM MECHANICS USING MAPLE V

Yutaka Abe
Quantum Instrumentation Laboratory, Hokkaido University, Sapporo, Japan

Abstract

This article discusses a set of Maple programs developed to reinforce understanding of basic concepts presented in an introductory quantum mechanics course. It was designed so that the solutions of the problems were analytically tractable as well as numerically obtainable. We note that there exist various approaches to a simple quantum mechanical problem. For example, for the problem of bound-states with a one-dimensional potential, the transfer matrix method, the Laplace transform method, and the Feynman path integral method can equally be applied. We believe that this type of program is extremely useful for igniting the student's interest and widening his or her viewpoint. Several examples, such as the solution of various one-dimensional potential problems and the solution of the one-dimensional Schroedinger wave equation, are discussed in order to indicate how constructive reinforcement is established.

Mathematical Computation with Maple V:
Ideas and Applications
Tom Lee, Editor
©1993 Birkhäuser Boston

1. INTRODUCTION

There is much talk about computer-aided instruction in the field of physics and engineering, but it seems that the computer as a learning device has provided little help for students trying to grasp the basic ideas in these fields. Of course, various packages for numerical computations are now available [1], and it is very easy for instructors to introduce these packages in their classes. In my experience, this type of instruction has never succeeded in extending a student's ability for a further understanding of physics. There are always certain barriers between a lecturer's analytical description on a blackboard and the output of numerical computations by a computer. Sometimes the result in a classroom has been destructive, in that the students understand less physics and get fewer computational skills. Actually, numerical computations cannot teach students real physics.

On the other hand, it is true that there are a few problems in physics that can be solved analytically. For example, if you wish to explore the behavior of an electron wave function under a certain general potential function, you may suddenly encounter severe difficulty if you stick to some analytical solution. Numerical methods can treat a much wider variety of interesting problems than can be handled by analytical methods.

The motivation of this paper is to explore new methods of instruction balanced between analytical and numerical approaches.

The author selected Maple V for symbolic (analytical) and numerical calculations [2] because of its excellent interface with users, distinguished power for mathematical operations, and suitability for students exploring creative programming with small computers. In the following sections, we treat several elementary quantum mechanical problems with this approach.

2. ONE-DIMENSIONAL FINITE POTENTIAL BARRIER PROBLEMS

The problems of one-dimensional finite potential barriers and wells are treated universally in introductory quantum mechanics textbooks. There are a number of ways to attack these problems. In the simple type of potential, the overall wave function is constructed out of pieces having the form of a general wave function by matching the wave function and its derivative at the discontinuities of the given potential function. The transfer matrix method is one commonly used method.

Let us consider the Schrodinger equation for the potential function in Fig. 1. The Schrodinger equation for regions I, II, III are:

$$\frac{d^2\phi}{dx^2}+\alpha^2 x=0 \quad (x<0 \ and \ x>a)$$
$$(2.1.a),$$

$$\frac{d^2\phi}{dx^2}-\beta^2 x=0 \quad (0<x<a), \ V>\varepsilon$$
$$(2.1.b),$$

$$\frac{d^2\phi}{dx^2}+\mu^2 x=0 \quad (0<x<a), \ V<\varepsilon$$
$$(2.1.c)$$

where

$$\alpha^2=2m\varepsilon/\hbar^2, \quad \beta^2=2m(V-\varepsilon)/\hbar^2$$

$$\mu^2=2m(\varepsilon-V)/\hbar^2$$
$$(2.2).$$

We take the wave functions in each region as:

$$\phi_I=\exp(i\alpha x)+r\exp(-i\alpha x), \quad x<0$$
$$(2.3.a)$$

$$\phi_{II}=A\exp(-\beta x)+B\exp(\beta x), \quad 0<x<a$$
$$(2.3.b)$$

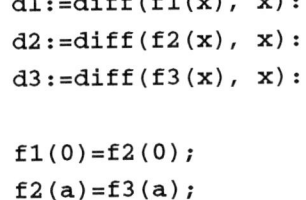

$$\phi_{II} = A\exp(-\beta x) + B\exp(\beta x), \quad 0 < x < a$$

(2.3.c).

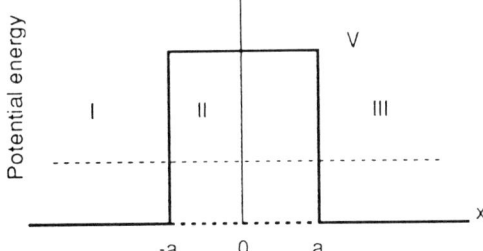

Fig. 1 One-dimensional
   square potential barrier.

Now let us try to find the
solutions for the above problem.

We obtain 4 linear equations from
the boundary condition at x =0 and
x=a. The solutions for A, B, T,
and R are :

```
• f1 := x -> exp(I*alpha*x) +
R*exp(-I*alpha*x):
  f2 := x -> A*exp(-beta*x) +
B*exp(beta*x) :
  f3 := x -> T*exp(I*alpha*x):
```

```
d1:=diff(f1(x), x):
d2:=diff(f2(x), x):
d3:=diff(f3(x), x):

f1(0)=f2(0);
f2(a)=f3(a);
```

```
evalf(subs(x=0,d1))=evalf(subs(x=0,
d2));

evalf(subs(x=a,d2))=evalf(subs(x=a,
d3));

1 + R = A + B

A exp(- beta a) + B exp(beta a) = T
exp(I alpha a)

1. I alpha - 1. R I alpha =  - 1. A
beta + 1. B beta

- 1. A beta exp( - 1. beta a) + B
beta exp(beta a) =

T I alpha exp(I alpha a)
```

We obtain 4 linear equations from
the boundary conditions at x=0 and
x=a. The solutions for A, B, T, R
are;

```
• e1:= 1 + R = A + B:
  e2:= I*alpha - R*I*alpha = -A*beta + B*beta:
  e3:= A*u + B*v = T*s:

  e4:= -A*beta*u + B*beta*v = T*I*alpha*s:
  solve({e1,e2,e3,e4},{R,T,A,B});
```

```
              v I alpha (I alpha - beta)
  {A = 2 ---------------------------,
                      %1


             (beta + I alpha) u I alpha
      B = - 2 ---------------------------,
                        %1


                u beta v I alpha
      T = - 4 ----------------,
                    s %1


              2         2           2         2
         - v alpha  - beta  v + alpha  u + beta  u
      R = ---------------------------------------}
                            %1


            2        2                             2
%1 := - v alpha  + beta  v - 2 I alpha u beta + alpha  u


           2
     - beta  u - 2 v I alpha beta
```

Here, u=exp(beta a), v= exp(beta a), and s = exp(i alpha a). Substituting these values into the expressions of T, R, A, and B, the amplitudes of the wave functions are completely determined. For example,

$$T = \frac{4i\alpha\beta u\, v}{s\{(\alpha^2-\beta^2)(u-v)+2i\alpha\beta(u+v)\}}$$

(Calculation of TT*)

$$= \frac{2i\alpha\beta}{\exp(i\alpha a)\{(\alpha^2-\beta^2)\sinh(\beta a)+2i\alpha\beta\cosh(\beta a)\}}$$

(2.4)

The transmission coefficient TT* is, therefore

$$TT^* = \frac{4\alpha^2\beta^2}{(\alpha^2+\beta^2)\sinh^2(\alpha a)+4\alpha^2\beta^2}$$

- T1:=2*I*alpha*beta*exp(-I*alpha*a)/((alpha^2 -
  beta^2)*sinh(beta*a) + 2*I*alpha*beta*cosh(beta*a));

```
                        I alpha beta exp(- I alpha a)
T1 := 2 -------------------------------------------------------
                2       2
          (alpha  - beta ) sinh(beta a) + 2 I alpha beta cosh(beta a)
```

- T2:=-2*I*alpha*beta*exp(-I*alpha*a)/((alpha^2 -
beta^2)*sinh(beta*a) - 2*I*alpha*beta*cosh(beta*a));

```
                          I alpha beta exp(- I alpha a)
T2 := - 2 -------------------------------------------------------
                2       2
          (alpha  - beta ) sinh(beta a) - 2 I alpha beta cosh(beta a)
```

- evalc(T1*T2):
simplify(");

```
     2      2                 2
4 beta  alpha  (2 cos(alpha a)  - 2 cos(alpha a) sin(alpha a) I - 1)/

       4              2       4         2      2             2
  (alpha  cosh(beta a)  - alpha  + 2 alpha  beta  cosh(beta a)

            2      2       4             2       4
     + 2 beta  alpha  + beta  cosh(beta a)  - beta )
```

Numerical Examples of the Transmission Coefficients

- m:=9.01*10^(-31):
 hbar:=1.0546*10^(-34):
 v:=0.5*1.602*10^(-19):
 en:= ep*1.602*10^(-19):
 alfa:=sqrt(2* m* en)/hbar:
 beta:=sqrt(2* m*(v-en))/hbar:
 a:=10*10^(-10):
t1:=4* beta^2* alfa^2:
t2:=((alfa^2 + beta^2))^2* (sinh(beta* a))^2 + 4* beta^2* alfa^2:
d:=t1/t2:

```
p:=simplify(d);
```

$$p := 4.\, ep\,(\,-.1684315858*10^{10} + .3368631715*10^{10}\ ep\,)\,/\,($$

$$-.842157929*10^{9}\ \cosh(3.602515719\,(1. - 2.\,ep)^{1/2})^{2} + .842157929*10^{9}$$

$$-.6737263430*10^{10}\ ep + .1347452686*10^{11}\ ep^{2}\,)$$

- `plot(log(p), ep=0.01..0.48);`

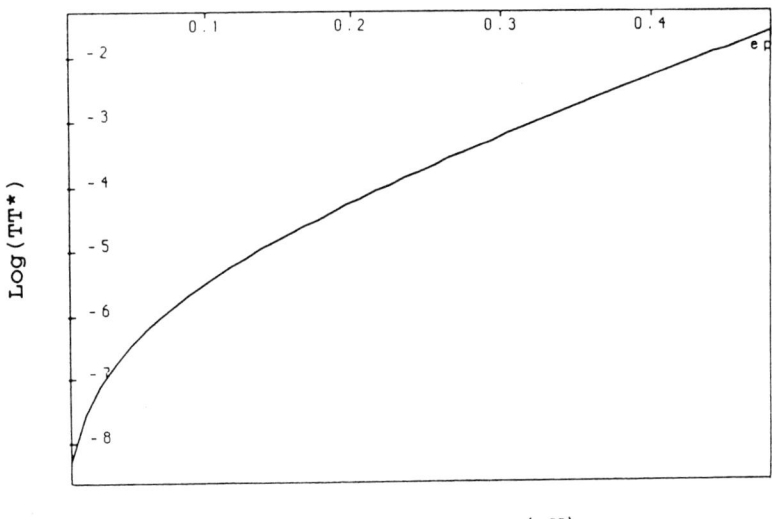

Fig. 2 Tansmission coefficient as a function of energy of incident electron. The width of the barrier 10 angstrom and the hight of the barrier is 0.5 eV.

We have illustrated the quantum mechanical tunneling phenomena using the symbolic manipulation program. It is seen that there is well-balanced combination of analytical as well as numerical approaches for this elementary problem.

Of course there are various interesting approaches to this potential problem. How can you treat a one-dimensional potential problem in the form of the potential is more general than the square barrier? By the WKB method [3] you might obtain rigorous solutions. Anyone can understand intuitively that the approximate solution for very general potential barrier can be obtained by a large number of segments of constant height, as shown in Fig. 3. Essentially this is the same concept that a definite integral of some function can be evaluated by summing the segments covered over the integral area.

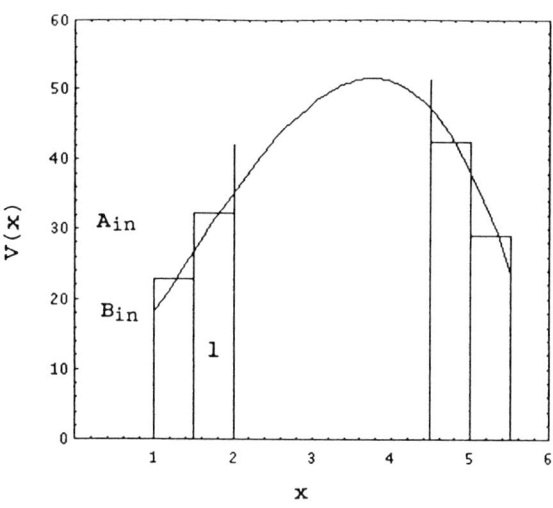

Fig. 3. Approximation of potential by segments of flat barrier.

Indeed, Kalotas and Lee[4] have reported the exact formulation of this general approach. We follow their suggestions in Maple V procedures.

The continuous potential V(x) is segmented into n discrete flat barriers each of width w and height V_i, where

$$V_i = \frac{1}{2}\{V[l - (i - 1)w] + V[l + iw]\},$$

$$i = 1, 2, .., n$$

(2.6).

For convenience, we assume all the Vi satisfy $V_i > E$. Defining k^2 and α^2 as

$$k^2 = 2mE/\hbar^2, \quad \alpha_i^2 = 2m(V_i - E)/\hbar^2$$

(2.7)

the solution of the Schroedinger equation within the barrier is

$$\psi_i = A_i \exp(\alpha_i x) + B_i \exp(-\alpha_i x)$$

(2.8).

On the other hand, the solutions in the regions outside the barrier are

$$\psi_{in} = A_{in} \exp(ikx) + B_{in} \exp(-ikx)$$

(2.9),

and

$$\psi_{out} = A_{out} \exp(ikx)$$

(2.10).

The boundary conditions of the wave function ant its derivative at x-1, l+w, l+2w, ..., l+ nw establishes a linear relation between the A and B coefficients. The relation can be expressed in a matrix form and this is called the **Transfer Matrix**. For example, ψ and ψ' at x=1 lead to

$A_{in} \exp(ikl) + B_{in} \exp(ikl) = A_1 \exp(\alpha_1 l) + B_1 \exp(-\alpha_1 l)$

$$(2.12b).$$

$ikA_{in} \exp(ikl) - ikB_{in} \exp(ikl) =$

$\alpha_1 A_1 \exp(\alpha_1 l) - \alpha_1 B_1 \exp(-\alpha_1 l)$

$$(2.11).$$

This can be expressed as

$$\begin{bmatrix} \exp(ikl) & \exp(-ikl) \\ ik\exp(ikl) & -ikl\exp(-ikl) \end{bmatrix} \begin{bmatrix} A_{in} \\ B_{in} \end{bmatrix} =$$

$$\begin{bmatrix} \exp(\alpha_1 l) & \exp(-\alpha_1 l) \\ \alpha_1 \exp(\alpha_1 l) & -\alpha_1 \exp(-\alpha_1 l) \end{bmatrix} \begin{bmatrix} A_{out} \\ B_{out} \end{bmatrix}$$

$$(2.12a),$$

$$M[l,ik]\begin{bmatrix} A_{in} \\ B_{in} \end{bmatrix} = M[l,\alpha_1]\begin{bmatrix} A_{out} \\ B_{out} \end{bmatrix}$$

The relation between A_{in}, B_{in}, and A_{out}, B_{out} is

$$\begin{bmatrix} A_{in} \\ B_{in} \end{bmatrix} = M^{-1}[l,ik]K[\alpha_1,w]K[\alpha_2,w]...K[\alpha_n,w]$$

$$\times M[l+nw,ik]\begin{bmatrix} A_{out} \\ B_{out} \end{bmatrix}$$

$$(2.13),$$

where

$$K[\alpha_i,w] = \begin{bmatrix} \cosh(\alpha_i w) & -(1/\alpha_i)\sinh(\alpha_i w) \\ -\alpha_i \sinh(\alpha_i w) & \cosh(\alpha_i w) \end{bmatrix}$$

$$(2.14).$$

Calculation of M[L,a1] M-1[1+w, a1]

• with (linalg):

• m1:=matrix([[exp(a1* 1), exp(-a1* 1)], [a1*exp(a1*1), -a1*exp(-a1*1)]]);

```
              [   exp(a1 1)        exp(- a1 1)    ]
       m1  := [                                   ]
              [ a1 exp(a1 1)    - a1 exp(- a1 1)  ]
```

m2:=matrix([[exp(a1*(1+w)), exp(-a1*(1+w))], [a1*exp(a1*(1+w)), -a1*exp(-a1*(1+w))]]);

```
                    [  exp(a1 (1 + w))       exp(- a1 (1 + w))    ]
          m2 := [                                                 ]
                    [ a1 exp(a1 (1 + w))  - a1 exp(- a1 (1 + w)) ]
```

- **m3:=inverse(m2);**

```
                [             1                          1            ]
                [  ----------------       --------------------        ]
                [  2 exp(a1 (1 + w))      2 exp(a1 (1 + w)) a1        ]
          m3 := [                                                     ]
                [             1                          1            ]
                [  ------------------    - --------------------       ]
                [  2 exp(- a1 (1 + w))    2 a1 exp(- a1 (1 + w))      ]
```

- **evalm(m1 &* m3);**

```
                            %2 + %1
          [1/2 --------------------------------,
              exp(a1 (1 + w)) exp(- a1 (1 + w))


                              %2 - %1
            1/2 -------------------------------------]
                exp(a1 (1 + w)) a1 exp(- a1 (1 + w))


                          a1 (%2 - %1)
          [1/2 --------------------------------,
              exp(a1 (1 + w)) exp(- a1 (1 + w))




                              %2 + %1
            1/2 -------------------------------]
                exp(a1 (1 + w)) exp(- a1 (1 + w))


      %1 :=                    exp(- a1 1) exp(a1 (1 + w))

      %2 :=                    exp(a1 1) exp(- a1 (1 + w))
```

If we denote the product of the k matrix as

$$K[\alpha_1,w]K[\alpha_2,w]\ldots\ldots K[\alpha_n,w] \equiv \begin{bmatrix} P & Q \\ R & S \end{bmatrix}$$

(2.15),

then the transmission coefficient TT* is given by

$$TT^* = \left|\frac{A_{out}}{B_{out}}\right|^2 = \frac{4}{[P+S)^2+(Qk-R/k)^2]}$$

(2.16).

When Vi< ε, a becomes imaginary, α=iβ , so that the assumption of Vi < ε can be relaxed to all the values of a particle energy.

Calculation of RR*

It is rather easily verified that Eq. (2.16) is reduced to Eq. (2.4) when the potential is a square flat barrier, and this offers a good exercise to the student.

Returning to the one-dimensional square potential problem again, let us study the situation where the incident electron energy is much larger than the potential energy. Since R has already been obtained in the preceding paragraph, we apply that formula for ε > V. The wave function in the barrier region with this condition, is given by

$$\psi_{II} = A\cos(\gamma a) + B\sin(\gamma a)$$

where

$$\gamma^2 = 2m(V-\varepsilon)/h^2 .$$

- R:=(alpha^2 -gamma^2)*sin(gamma*a)/((alpha^2 + gamma^2)*sin(gamma*a) + 2*I*alpha*gamma*cos(gamma*a));

$$R := \frac{(alpha^2 - gamma^2)\ sin(gamma\ a)}{(alpha^2 + gamma^2)\ sin(gamma\ a) + 2\ I\ alpha\ gamma\ cos(gamma\ a)}$$

- RC:=(alpha^2 - gamma^2)*sin(gamma*a)/((alpha^2 + gamma^2)*sin(gamma*a) - 2*I*alpha*gamma*cos(gamma*a));

$$RC := \frac{(alpha^2 - gamma^2)\ sin(gamma\ a)}{(alpha^2 + gamma^2)\ sin(gamma\ a) - 2\ I\ alpha\ gamma\ cos(gamma\ a)}$$

- **simplify(expand(R*RC));**

$$(- \alpha^4 + \alpha^4 \cos(\gamma a)^2 + 2 \alpha^2 \gamma^2$$

$$- 2 \alpha^2 \gamma^2 \cos(\gamma a)^2 - \gamma^4 + \gamma^4 \cos(\gamma a)^2) \Big/ \Big/$$

$$(- \alpha^4 + \alpha^4 \cos(\gamma a)^2 - 2 \alpha^2 \gamma^2$$

$$- 2 \alpha^2 \gamma^2 \cos(\gamma a)^2 - \gamma^4 + \gamma^4 \cos(\gamma a)^2)$$

- factor(");

$$(- 1 + \cos(\gamma a)) (\cos(\gamma a) + 1) (\alpha - \gamma)^2 (\alpha + \gamma)^2$$

$$\Big/ \Big/ ((\cos(\gamma a) \alpha^2 - \cos(\gamma a) \gamma^2 - \alpha^2 - \gamma^2)$$

$$(\cos(\gamma a) \alpha^2 - \cos(\gamma a) \gamma^2 + \alpha^2 + \gamma^2))$$

From the above calculations, RR* is expressed as

$$RR^* = \frac{V^2 \sin^2(\gamma a)}{V^2 \sin^2(\gamma a) + 4\varepsilon(\varepsilon - V)}$$

(2.17).

From Eq.(2.17) it is easily seen that RR* approaches zero when the electron energy is large enough compared with V. However, due to sin2(ga) factor involved in Eq.(2.17), there is resonant scattering of electrons from the potential barrier.

Numericalexample of the resonant scattering.

```
• m:=9.01*10^(-31):
hbar:=1.0546*10^(-34):
v:=0.5*1.602*10^(-19):
en:=ep*1.602*10^(-19):
g:=sqrt(2*m*(en-v))/hbar:
a:=10*10^(-10):
p1:=v^2*(sin(g*a))^2:
p2:=v^2*(sin(g*a))^2 + 4*en*(en-v)
r:=p1/p2:
plot(log(r), ep=1..20,
numpoints=500);
```

It is seen from Fig. 4 that there are resonant transmission of incident electrons when ba=n pi. We can expect that similar resonant transmission can be observed in a one-dimensional square well potential.

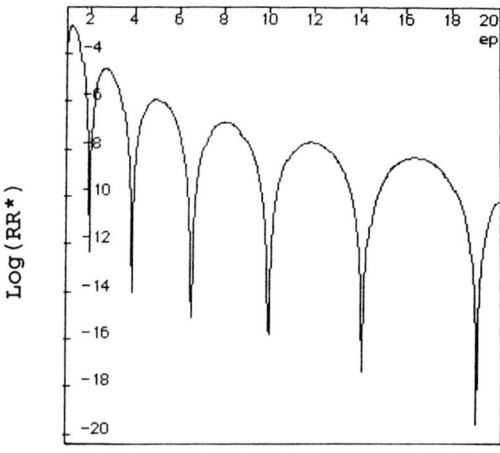

electron energy (eV)

Fig.4. Numerical example of resonant scattering by one-dimentional square barrier.

We have shown several examples of how to use Maple V for simple one-dimensional potential problems. Some of the results can be easily extended to solve the electronic states in one-dimensional periodic potential arrays [5], which is the basis of the general theory of band structures in solids. The Laplace transformation method [6] is one of the more powerful approaches to periodic potential problems.

A weakly coupled harmonic oscillator is one interesting problem, from which we can introduce the model of lattice vibrations in a simple crystal. Feynman's path integral is another approach to the tunneling problem [7]. Although it is necessary to give a comprehensive introduction to the physical meaning involved in Feynman's path integral, we successfully introduce this approach using Maple V in the class of senior undergraduate students.

3. Time-Dependent Solutions of Wave Packets

We have shown several typical examples of stationary state wave functions in simple potentials in the previous sections. The time-dependent behaviors of the wave functions are one of most difficult problems for the undergraduate students to grasp their physical meaning.

Here, we illustrate some of the visualized instruction for this kind of **problem**.

In order to describe a particle at a local position in quantum mechanics, it is necessary to construct a wave packet. A free electron wave function with momentum p is

$$\phi_p = (2\pi\hbar)^{-1/2} \exp\left[\frac{-i(\varepsilon t - px)}{\hbar}\right]$$

$$(3.1a),$$

which satisfies the time-dependent Schroedinger equation

$$i\hbar\frac{\partial}{\partial t}\phi_p = -\frac{\hbar^2}{2m}\frac{\partial^2}{\partial x^2}\phi_p$$

$$(3.1b).$$

Since the Eq.(3.1b) is linear, the superposition

$$\psi(x,t) = \phi_p \sum_{n=1}^{N} a_n \phi_{pn}(x,t)$$

$$(3.2a)$$

is also the solution of the wave equation, where pn corresponds to different momenta pn. Replacing the sum by an integral, we have

$$\psi(x,t) = \int_{-\infty}^{\infty} f(p)\phi_{pn}(x-x_0,t)dp$$

$$(3.2b),$$

where f(p) is the weighting function for the different momenta p. Let us consider the case where f(p) is given by a Gaussian distribution function:

$$f(p) = \frac{1}{(2\pi)^{1/4}\sqrt{\sigma_p}} \exp\left[-\frac{(p-p_0)^2}{4\sigma_p^2}\right]$$

$$(3.3).$$

Using the Fourier transformation, the probability density for observing the free particle at position x and time t is given by

$$|\psi(x,t)|^2 = \frac{1}{\sqrt{2\pi}\sigma_x} \exp\left[\frac{[x-(x_0+vt)^2}{2\sigma_x^2}\right]$$

$$(3.4a),$$

15

where

$$\sigma_x^2 = \sigma_{x0}^2 + \frac{\hbar^2 t^2}{4\sigma_{x0}^2}, \quad \sigma_{x0} = \frac{\hbar}{2\sigma_p}$$

(3.4.b).

A very important point about the wavepacket is the fact that it must satisfy **Heisenberg's Uncertainty Principle.** In fact, the widths σ_x and σ_p satisfy the relation

$$\sigma_x \, \sigma_p \geq \hbar/2$$

Examples of the wavepacket dispersion and the propagation of wavepacket are illustrated in Fig. 5a and Fig. 5b.

It is easily realized from Fig 5a and 5b that the wavepacket behaves as a probability density in a diffusion process. Actually if we put $\tau = it/h$ in the time-dependent Schroedinger equation, we have

$$\frac{\partial \psi(x,\tau)}{\partial \tau} = \frac{\hbar^2}{2m} \frac{\partial^2 \psi(x,\tau)}{\partial x^2}$$

(3.5a),

and this is a typical diffusion equation with a diffusion constant D, where

$$D = \frac{\hbar^2}{2m}$$

(3.5b).

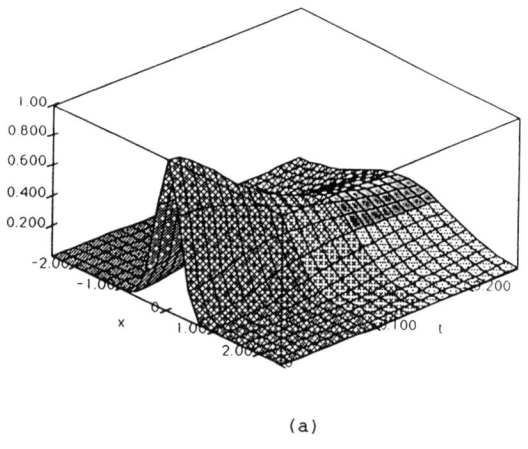

(a)

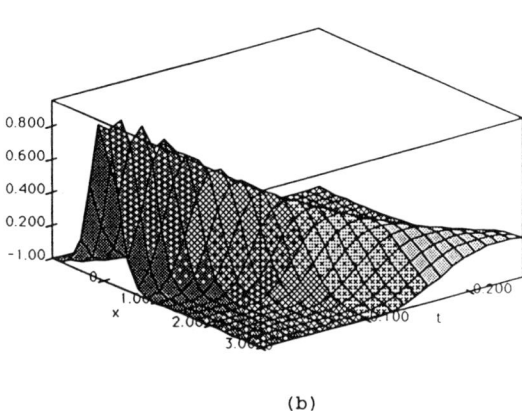

(b)

Fig. 5. (a) Diffusion of a Gaussian wavepacket.
    (b) Propagation and diffusion of a Gaussian wavepacket.

It is noteworthy that the Schroedinger wave equation takes very different mathematical form from a classical wave equation. It is said that the concept of the imaginary time domain was first suggested by E. Fermi. Various textbooks of mathematical physics present complete solutions for diffusion equation, most of which are concerned with the problem of thermal conduction in an infinite or finite medium.

The diffusion processes is considered to be random walk processes, and therefore the Monte Carlo method is naturally introduced in this phenomena. In Fig.6 , numerical results of Monte Carlo simulations on the ground state wave function of simple harmonic oscillator are shown.

It is rather easy to extend the above discussion into the problem of wavepacket propagation through a simple potential barrier or well, and this offers a very good exercise for the students to learn about a scattering phenomenon of the wavepacket.

The most simple computational technique for this problem is the finite difference method in which the wave equation is replaced by the corresponding difference equation:

$$i\hbar\frac{\partial\psi(x_i,t)}{\partial t}\approx$$

$$-\frac{\hbar^2}{2m}\frac{\psi(x_{i+1},t)-2\psi(x_i,t)+\psi(x_{i-1},t)}{\delta x^2}+V(x_i,t)\psi(x_i,t)$$

$$(3.6).$$

By performing the actual computational procedures, the students will realize the precision and the limitation of the finite difference method as well as the the dynamical behavior of the wavepacket, namely the transmission and the reflection of the wavepacket (Fig. 7).

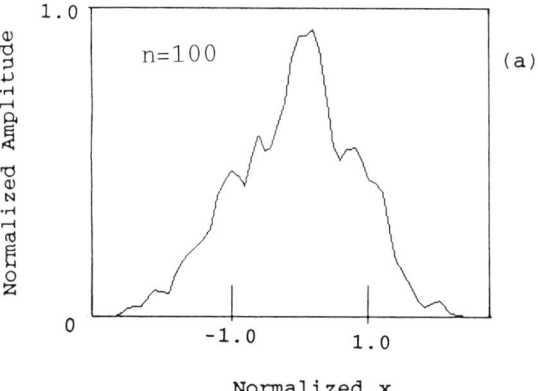

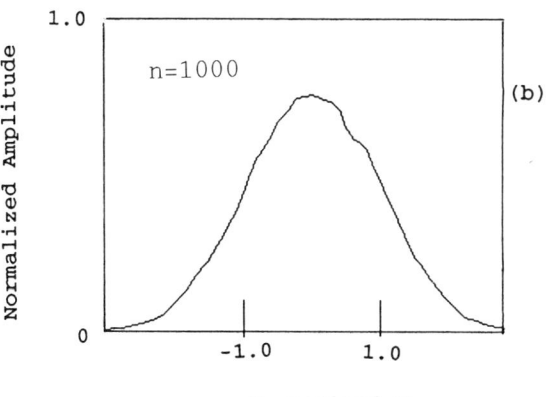

Fig. 6. Monte Carlo Simulation of the ground-state of simple harmonic oscillator.

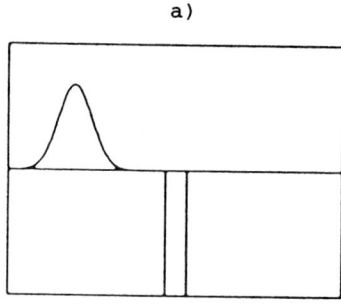

a)

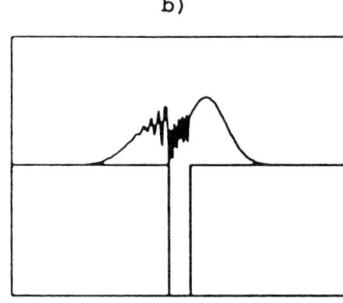

b)

Fig. 7. Transmission and reflection of a Gaussian wavepacket by square potential well. a) t=0, b) t=300.

5. Concluding Remarks

We have shown several well-balanced computational programs for elementary quantum mechanics which involve analytical as well as numerical calculations using the symbolic manipulation programs, Maple V. We would like to point out the following remarks for this program:

1) It is not necessary to provide all the topics which are usually described in the standard textbooks. A few basic problems are enough for the students to grasp the basic ideas involved in quantum mechanics. The most important point for the success of the program is to illustrate to the students that there exist various approaches to the same problem, rather than just to show a collection of numerical examples.

2) It is very dangerous for the instructor to overestimate the effect of numerical results on the basic understanding of physics. The instructors should endeavor to seek creative computer programs which involve well-balanced analytical and numerical solutions. These programs may continue to keep the student's interest and we can expect these kinds of programs to reinforce the student's basic understanding of quantum mechanics.

3) The instructors should illustrate the fact that there exists basic principles which can be applied for a wide variety of the problems.

We have concretely shown that Maple V offers excellent tools to satisfy the above mentioned facts.

References

[1] There are published a number of packages. Among these packages one of the typical numerical calculations is:
W. H. Press et al, "Numerical Recipes-The Art of Scientific Computing",(Cambridge Press, Cambridge 1986).

[2] B. W. Char t al,"Maple V Language Reference Manual",(Springer-Verlag, 1991).

[3]E. Merzbacher, "Quantum Mechanics", Chapt. 7, (John Wiley & Sons, New York 1961).

[4] T. M. Kaltos and A. R. Lee, "A new approach to one-dimensional scattering", Am.J. Physics, 59 (1990) 48-52.

[5] Y. Abe,"Bound-State Solutions for One-Dimensional Periodic Potential with Multiple Barriers per Period, Bulletin of the Faculty of Engineering, Hokkaido University, 158, (1992) 57-69.

[6] P. H. A. Santana and A. Rosato, "Use of the Laplace transformation to solve the one-dimensional periodic potential problem", Am. J. Phys. 41, (1973) 1138-1144.

[7] B. R. Holstein and A. R. Swift, "Barrier penetration via path integral", Am. J. Phys. 50, (1982) 833-839.

Yutaka Abe

Professor Abe received a BS degree in electrical engineering from Hokkaido University (Japan) in 1954. He joined Electrotechnical Laboratories in 1956 and worked on basic physics of semiconductor devices. In 1966, he received a doctoral degree in applied physics from Hokkaido University and, in 1967 joined their department of applied physics. He was appointed a full Professor of Engineering at Hokkaido University in 1973. His personal fax number is 011-757-1162 and his email address is ya@hune.hokudai.ac.jp

COMBINATORIAL OBJECTS AND THEIR GENERATING FUNCTIONS: A MAPLE CLASS ROOM ENVIRONMENT

John.S. Devitt
Department of Combinatorics and Optimization, University of Waterloo
(on leave from the University of Saskatchewan), Waterloo ON, Canada

1 Introduction

Combinatorial generating functions are an important part of any introduction to combinatorics. A good introduction to them must reenforce the concepts of weight functions and the existence of a one to one correspondence between the monomials representing the objects in the combinatorial set and the objects themselves. An entire set is represented by a sum of monomials. The combinatorial information about the set is deduced by simplifying or manipulating that sum.

An effective computational environment for studying this topic must include facilities for the generation and manipulation of a large variety of sets of combinatorial objects and the corresponding generating functions. There must also be control over the amount of user interaction required to complete each transformation.

While considerable support for generating combinatorial objects already exists in Maple through the use of lists, sets and through various routines in packages such as *combinat* and *numtheory* and a variety of representations of generating functions by way of packages such as *genfunc*, *powseries* and other generating function packages found in the share library none of these address two important questions:

1. The automatic construction and manipulation of generating functions using one or more weight functions simultaneously.

2. The extension of these tools to infinite combinatorial sets often arising in connection with the representation of generating functions in closed form.

Mathematical Computation with Maple V:
Ideas and Applications
Tom Lee, Editor
©1993 Birkhäuser Boston

In this paper we introduce a package that is specially designed to address these two key points and which is in use across all introductory combinatorics sections at the University of Waterloo. The package includes support for the representation of infinite sets, the systematic generation of portions of those sets, and for obtaining multivariate generating functions in closed form. There are a wide variety of specialized operations for binomial series and their combinations.

We begin with several examples of how the package can be used to solve enumeration problems for sequences over a finite alphabet. This is followed by a discussion of the key issues that arise in attempting to deal with these types of problems and some concluding remarks.

2 Examples

We focus primarily on sequences over a finite alphabet though other combinatorial objects can be supported using the same tools. The class of sequences we are concerned with are those constructed by regular expressions.

Example 2.1 *Generate the collection S of all sequences over the alphabet $\{0, 1, 2\}$.*

Solution: For example, the collection S of all sequences over the alphabet $\{0, 1, 2\}$ can be generated as

```
> A := alphabet(0,1,2);
            A := {0, 1, 2}

> Star({0,1,2});
                        *
            [{0, 1, 2}]

> S := genseq(");
S := [phi, 0, 1, 2, 00, 10, 20, 01, 11,

    21, 02, 12, 22, 000, 100, 200, 010,

    110, 210, 020, 120, 220, 001, 101,

    201, 011, 111, 211, 021, 121, 221,

    002, 102, 202, 012, 112, 212, 022,
```

```
                122, 222] + Objects(nops = 4)
```

This set is infinite

```
> cardinality(S);
                        infinity
```

so only some of the sequences can be displayed. They are chosen to be of minimum weight according to a user definable weight function. By default, at least 30 are displayed and the default weight function is **nops( )**.

The displayed sequences can be extracted as a set or a list from the above collection by use of the **select( )** command.

```
> displayed_sequences(S);
  [phi, 0, 1, 2, 00, 10, 20, 01, 11, 21,

    02, 12, 22, 000, 100, 200, 010,

    110, 210, 020, 120, 220, 001, 101,

    201, 011, 111, 211, 021, 121, 221,

    002, 102, 202, 012, 112, 212, 022,

    122, 222]
```

For example, those having at least one occurrence of the sequence 01 are found by the command

```
> select( countsubstring[0,1]>0, S );
    [01, 010, 001, 101, 201, 011, 012]
```

The **genfunc( )** command works with either the result produced by a call to **genseq( )**, sets of sequences or regular expressions . Lists are used denote sequence concatenation as in

```
> T := genseq([A,A]);
T :=

    [00, 10, 20, 01, 11, 21, 02, 12, 22]
```

The sequences produced in this manner form a monoid with coefficients chosen from a ring as in

```
> genseq(T  + u*Seq(0,0) + 3*Seq());
  [3 phi, (u + 1) 00, 10, 20, 01, 11, 21,

    02, 12, 22]
```

Example 2.2 *Find the average number of blocks in the set of all 0-1 sequences of length n.*

Solution:
A block is a maximal string of 0's or a maximal string of 1's. The set of all such strings is generated by the pattern

```
> alphabet(0,1);
                    {0, 1}

> pat := [Star(0),
>    Star( 1,Star(1),0,Star(0)),
>    Star(1)];
               *       *      * *   *
     pat := [0 , [1, 1 , 0, 0 ] , 1 ]
```

The generating function based on this pattern with respect to length, number of 0's and number of 1's is

```
> alias( w=countsubstring ):
> genfunc(pat,[t,x,y],[nops,w[0],w[1]]);
           /                  2            \
           |                 t  y x        |
 1/((1 - t x) |1 - ------------------|
           \        (1 - t y) (1 - t x)/

      (1 - t y))
```

We still need to keep track of blocks. To do so we introduce an indeterminant into the pattern to record every occurrence of every block.

```
> pat2 := [ [phi + u*[0,Star(0)]] ,
> Star(u*phi,1,Star(1),phi*u,0,Star(0)),
> [phi+u*[1,Star(1)]]];
                                *
    pat2 := [[phi + u [0, 0 ]],

                    *            * *
        [u phi, 1, 1 , u phi, 0, 0 ] ,

                        *
        [phi + u [1, 1 ]]]
```

Here, $phi = \phi$ denotes the null string. The sequences produced by this pattern and displaying at least the first 50 are

```
> S := genseq(pat2,nops, 50 );
                                   2
S := [phi, u 1, u 0, u 00, u  10,

      2         2       3         2        3
    u  01, u  11, u 000, u  100, u  010,

      2        2       3         2
    u  110, u  001, u 101, u  011,

      3        3         3
    u 111, u 1101, u  1001, u  1011,

      2         2         2
    u  1000, u  1100, u  1110, u 0000,

      4         4         3
    u 1111, u  1010, u  0101, u  0100,

      3         2         2         3
    u  0110, u  0001, u  0011, u  0010,

      2         2         2
    u  0111, u  00001, u  00011,

      3         4         2
    u  00110, u  00101, u  00111,

      3          3         3
    u  11101, u  11001, u  10001,

      3          3         3
    u  11011, u  10011, u  10111,

      3          2         2
    u  00100, u  11110, u  11000,

      2          5         2
    u  10000, u  10101, u  11100,

      3
    u  00010, u 00000, u 11111,

      4          4         4
    u  10100, u  10110, u  10010,

      4          2         3
    u  11010, u  01111, u  01110,
```

```
       3         3         4
    u  01100, u  01000, u  01101,

       4         4         5
    u  01011, u  01001, u  01010]

    + Objects(nops = 6)
```

The corresponding generating function with respect to length of the sequence is

```
> genfunc(S,x,nops);
                 2   2
    1 + 2 u x + (2 u + 2 u ) x

         2      3     3
    + (2 u + 4 u  + 2 u ) x

         3      2        4     4
    + (6 u + 6 u  + 2 u + 2 u ) x

         2       3      4     5        5
    + (8 u + 12 u  + 8 u + 2 u + 2 u) x

         6
    + O(x )
```

In closed form this is

```
> f := genfunc(S,x,closedform);
              /     u x \2
              |1 + -----|
              \    1 - x/
       f := -------------
                   2   2
                  u  x
              1 - --------
                        2
                  (1 - x)
```

which simplifies to

```
> f := normal(f):
```

To compute the average number of blocks we must be able to extract coefficients. To get the coefficient of x^n in the series corresponding to f we do

```
> f1 := convert(f,Series,x);
                infinity
                 -----
                  \             n1    n1
                   \         u (u + 1)  x
       f1 :=       )      2 ---------------
                  /             u + 1
                 -----
                n1 = 0

> c1 := getcoeff(",x,n);
                                n
                      u (u + 1)
            c1 := 2 ----------
                        u + 1
```

which simplifies to 2^n.

```
> subs(u=1,");
                     n
                    2
```

To compute the average number of blocks of size n we need to compute

```
> diff(c1,u)/2^n;
                   n                     n
             (u + 1)          u (u + 1)
         (2 --------- - 2 -----------
             u + 1                2
                           (u + 1)

                       n
            u (u + 1)   n      / n
         + 2 -----------)     /  2
                     2       /
              (u + 1)

> subs(u=1,x=1,");
              n           n
         1/2 2  + 1/2 2  n
         -----------------
                 n
                2
```

We can also obtain this as

```
> diff(f,u);
               x              (1 - x + u x) x
        - ----------- + ----------------
           u x + x - 1                  2
                            (u x + x - 1)

> g := subs(u=1,");
                  x             x
        g := - ------- + -----------
               2 x - 1             2
                          (2 x - 1)

> convert(g,Series,x);
            infinity
             -----
              \             n1            n1    n1
               )      (1/2 2   n1 + 1/2 2  ) x
              /
             -----
            n1 = 1

> c2 := getcoeff(",x,n);
                    n          n
          c2 := 1/2 2  + 1/2 2  n
```

so that the average number of blocks is just

```
> normal( subs(u=1,c2/c1) );
              1/2 + 1/2 n
```

The specialized series represented by *Series* can be manipulated in a manner similar to Maple's ordinary series data structure. Consider the two series

```
> s1 := convert( 1/(1-x)^2,Series );
            infinity
             -----
              \             n1 [ -2 ] n1
        s1 :=  )       (-1)   [    ] x
              /                [ n1 ]
             -----
            n1 = 0

> s2 := convert( 1/(1+x)^3,Series );
            infinity
             -----
              \         [ -3 ] n1
        s2 :=  )        [    ] x
              /         [ n1 ]
             -----
            n1 = 0
```

These can be added, or multiplied as in

```
> evals( s1 + s2);
  infinity
  -----
   \       /     n1 [ -2 ]   [ -3 ]\ n1
    )     |(-1)   [    ] + [    ]| x
   /       \     [ n1 ]   [ n1 ]/
  -----
  n1 = 0
```

```
> expand(");
    /infinity                      \
    | -----                        |
    |  \            n1 [ -2 ] n1|
    |   )     (-1)    [    ] x   |
    |  /              [ n1 ]     |
    | -----                      |
    \ n1 = 0                      /

        /infinity                 \
        | -----                   |
        |  \          [ -3 ] n1|
      + |   )          [    ] x  |
        |  /           [ n1 ]    |
        | -----                  |
        \ n1 = 0                 /
```

or

```
> evals( s1*s2 );
  infinity
  -----
   \
    )          Sum(
   /
  -----
  n1 = 0

          n1_1 [  -2  ] [  -3  ]
    (-1)      [     ] [     ],
          [ n1_1 ] [ n1_2 ]

                              n1
    n1_1 + n1_2 = n1) x
```

and we can extract coefficients as in

```
> getcoeff(",x,k);
            n1_1 [  -2  ] [  -3  ]
    Sum((-1)    [     ] [     ],
            [ n1_1 ] [ n1_2 ]

    n1_1 + n1_2 = k)
```

simplify them

```
> simplify(");
  -----
   \                    (2 n1_1 + n1_2)
    )          (-1)
   /
  -----
  (n1_1 + n1_2) = k

              [ 2 + n1_2 ]
    (1 + n1_1) [          ]
              [   n1_2   ]
```

or find the value of them

```
> value(");
            (2 k)                        2
  1/16 (-1)      (15 + 14 k + 2 k

                                    2
    + 12 (k + 1) k + 4 (k + 1) k

               2            2
    - 14 (k + 1)  - 8 (k + 1) k
```

```
                 3
    + 4 (k + 1) )

                 (k - 1)               2
    - 1/16 (-1)         (11 + 10 k + 2 k )
```

This enables us to solve the following sorts of problems.

Example 2.3 *Find the coefficient of x^n in f defined by*

```
> f := (x+1)/( 2*x^2-3*x+1 );
              x + 1
    f := --------------
               2
          2 x  - 3 x + 1
```

Solution: This rational function is equivalent to

```
> s := convert(f,parfrac,x);
             3           2
    s := - ------- + -----
           2 x - 1   x - 1
```

which is a sum of two terms, each of which can be expanded using the binomial series. The corresponding series representation based on formulas for the nth coefficient are

```
> t := map(convert,s,Series,x);
          /infinity           \
          | -----             |
          |  \        n1 n1|
    t := |   )     3 2   x   |
          |  /                |
          | -----             |
          \ n1 = 0            /

          /infinity           \
          | -----             |
          |  \          n1|
        + |   )     - 2 x   |
          |  /              |
          | -----           |
          \ n1 = 0          /
```

The required coefficient is just

```
> map( getcoeff,",x,n);
       n
    3 2  - 2
```

The next example illustrates how the environment can be used to experiment with a given problem.

Example 2.4 *Find the generating function with respect to length for the sequences over the alphabet $\{0,1\}$ in which no substring of the form 0110 occurs.*

Solution: The set of all 0-1 sequences is generated by

```
> alphabet(0,1);
            {0, 1}
```

```
> S := genseq(Star("));
```

```
S := [phi, 0, 1, 01, 11, 00, 10, 010,

    110, 000, 100, 011, 111, 001, 101,

    0100, 1100, 0000, 1000, 0110, 1110,

    0010, 1010, 0101, 1101, 0001, 1001,

    0111, 1111, 0011, 1011]

    + Objects(nops = 5)
```

The displayed sequences that do not have 0110 are

```
> select(Not(hassubstring[0,1,1,0]),S);
[phi, 0, 1, 01, 11, 00, 10, 010, 110,

    000, 100, 011, 111, 001, 101, 0100,

    1100, 0000, 1000, 1110, 0010, 1010,

    0101, 1101, 0001, 1001, 0111, 1111,

    0011, 1011]
```

We can guess at the generating function by using these sample coefficients.

```
> genfunc( convert(",set) , x , nops );
         3     2      4
      8 x + 4 x + 15 x + 1 + 2 x

> closedform(");
closedform:
You may want to compare this estimate
with the next order of approximation!
              1
           ---------
            - 2 x + 1
```

This is the generating function for all 0-1 sequences so it is clearly not correct. The "guess" at the closed form for the generating function is wrong. However, we try again with a larger sample of coefficients.

```
> S := genseq(S,nops,100):
> select(Not(hassubstring[0,1,1,0]),S):
> genfunc( convert(",set) , x , nops );
      3     2      4      5      6
    8 x + 4 x + 15 x + 28 x + 52 x + 1

      + 2 x

> closedform(");
closedform:
You may want to compare this estimate
with the next order of approximation!
                3
         1 + 1/2 x
      ----------------
                     3
      1 - 2 x + 1/2 x
```

This guess is still wrong for it has fractional coefficients as in

```
> taylor(",x,10);
               2      3      4      5
    1 + 2 x + 4 x + 8 x + 15 x + 28 x

            6        7        8
       + 52 x + 193/2 x + 179 x

            9        10
       + 332 x + O(x  )
```

The right answer is produced by looking at over 300 sequences as in

```
> S := genseq(S,nops,300):
> select(Not(hassubstring[0,1,1,0]),S):
> genfunc( convert(",set) , x , nops );
      3      6      4      5       8
    8 x + 52 x + 15 x + 28 x + 181 x

           2      7
      + 4 x + 97 x + 1 + 2 x
```

The commands **genfunc( )** and **genseq( )** interact in such a way that all coefficients that are displayed are correct. In this case all sequences up to length 8 have been considered.

```
> closedform(");
closedform:
You may want to compare this estimate
with the next order of approximation!
                  3
            1 + x
      ------------------
              3      4
      1 + x  - 2 x - x
```

This is easily verified by using a regular expression to explicitly construct the same set of sequences.

3 Overview

The primary purpose of these tools is to provide support for interactive experimentation on sequence enumeration problems. A number of technical issues had to be confronted. In particular,

1. Any sequence is a concatenation of alphabet symbols. These symbols must remain distinct and recognizable if we are to effectively manipulate them.

2. The sets of sequences constructed by way of these regular expressions are typically infinite in size.

3. A naive cartesian product between two even modestly sized sets results in objects that are too large to manipulate or to deal with visually.

4. In order to manipulate and extract coefficients a representation that retained information about the computation of the nth coefficient was necessary.

Though they print as if concatenated they are actually represented as by an unevaluated function call. They are constructed by the command **Seq( )** as in

```
> alphabet(0,1,2,10);
           {0, 1, 2, 10}

> Seq(1,0);
              10

> Seq(1,10);
              1_10
```

and are actually objects such as

```
> lprint(");
_sequence(1,10)
```

A typical infinite set is generated by an expression like $\{0,1\}^*$. We call this the "star" product and enter it as

```
> Star({0,1});
                 *
           [{0, 1}]
```

Sequence concatenation is denoted by a list as in

```
> [Seq(2),",Seq(2)];
                    *
          [2, [{0, 1}] , 2]
```

in which each element is either an alphabet symbol or a regular expression or a set of sequences or alphabet symbols. We can also represent unions as '+' as in

```
> A := [ Seq(0,1,2,1,0) ];
          A := [01210]

> B := [ Seq(2,1,0,2,2) ];
          B := [21022]

> S := genseq(3*A + B);
          S := [3 01210, 21022]
```

The result is actually an multiset and it is somewhat analogous to Maple's truncated series objects.

```
> lprint(S);
_Genseq([[3, _sequence(0,1,2,1,0)], [1,
_sequence(2,1,0,2,2)]],nops,30,false,5,2
,3*_sequence(0,1,2,1,0)+_sequence(2,1,0,
2,2))
```

Typically only a sample of the sequences are generated and displayed. This is the first component of the data structure and lists some of the sequences along with their multiplicity (or more generally their coefficient from the sequence monoid). The remaining information is used to keep track of, for example, which weight function was in use, whether or not there are sequences in the set which are not displayed, the size of the set, the weight function used to sort the sequences, and finally a regular expression which can be used to regenerate the set if a larger sample of sequences is requested.

The concatenation of two sets as in

```
> A := Star(0);
                 *
            A := 0

> B := Star(1);
```

```
                 *
      B := 1
> S := genseq([A,B],nops,15);
 S := [phi, 0, 1, 00, 11, 01, 000, 111,

    011, 001, 0111, 0000, 1111, 0011,

    0001, 00000, 11111, 01111, 00111,

    00011, 00001] + Objects(nops = 6)
```

takes full advantage of the fact that a truncated representation of these infinite sets is in use and that the elements are sorted.

The routine **genfunc( )** works with either regular expressions or the combinatorial sets produced by **genseq( )**. The regular expressions are used to construct a closed form if a closed form is requested for a combinatorial set provided that such information is available.

```
> genfunc( S , x, nops , closedform);
               1
            --------
                 2
            (1 - x)
```

If the argument is an infinite set and no closed form has been requested then an explicit count of the sequences of the various weights is made and the truncated series corresponding to these sample sequences is given. The command **closedform( )** simply used pade approximates to guess at the closed form.

```
> genfunc(S,x,nops);
              2     3     4      5
   1 + 2 x + 3 x + 4 x + 5 x + 6 x

             6
        + O(x )

> closedform(");
closedform:
You may want to compare this estimate
with the next order of approximation!
            1
         -----------
                  2
         1 - 2 x + x
```

We also support multivariate generating functions. The generating function is regarded as a formal univariate series in the first variable.

```
> alias(w=countsubstring):
> genfunc(S,[x,s,t],[nops,w[0],w[1]]);
                    2    2
 1 + (s + t) x + (s + t + s t) x

        3    3    2      2      3
   + (s + t + s t + s  t) x

          3    4    4    2 2    3      4
   + (s t + s + t + s  t + s  t) x

   +

     5    5    4      2 3    3 2      4
   (s + t + s t + s  t + s  t + s  t)
```

```
  5      6
x   + O(x )
```

Maple's usual commands can be used as weight functions as well as user defined commands.

The select command has been greatly enhanced to facilitate a wide variety of selections based on weight. For example, the user may specify a selection of the form

```
> select(nops=2..4,S );
  [00, 11, 01, 000, 111, 011, 001, 0111,

     0000, 1111, 0011, 0001]
```

or

```
> select(nops> 8,S);
                     []
```

In addition, some indexed families of functions have been provided to aid with selections and as weight functions. We have seen examples of **countsubstring( )**. Another useful one is

```
> select( hassubstring[0,0,1,1] , S );
        [0011, 00111, 00011]
```

or

```
> select( Not(hassubstring[0]) , S );
     [phi, 1, 11, 111, 1111, 11111]
```

Where possible, the extendable Maple commands have been used to implement these objects and to facilitate interaction with these new objects. For example, each of **series( )**, **diff( )** and **value( )**, have enhancements.

4   Concluding Remarks

This package arose primarily as a vehicle to deliver tutorials for the introductory combinatorics course taught at the University of Waterloo. This course is taught at the second year level and is approximately half generating functions and half graph theory. It is based on departmental notes that have been refined over the past decade. Review material and assignments on enumeration and graph theory have been constructed using Maple V release 2 worksheets and now form part of that course. The objective is to give students a chance to discover the close correspondences between algebraic and set theoretic operations and to apply them to substantial problems.

The extent to which we have been able to accommodate all of the techniques dealt with in that course has been pleasantly surprising. The level of support this package provides for the basic operations provides a truly exciting way to experiment with combinatorial sets of a size that begins to show the general patterns that the theory predicts. This inability to consider fairly large examples has traditionally been a stumbling block for students in this course.

The author thanks both the Symbolic Computation Group and the Combinatorics Department at the University of Waterloo for the opportunity to explore this use of algebraic computing in a new environment.

Biography

J. S. Devitt completed his Masters degree in Combinatorial Number Theory under the supervision of Richard Guy at the University of Calgary in 1976. His Ph.D. in Combinatorics and Optimization was completed under the supervision of David Jackson at the University of Waterloo in 1981. He has been on staff at the University of Saskatchewan since 1984 where he is currently an associate professor of Mathematics. He is author of the recent text *Calculus with Maple V* and has had extensive experience in adapting Maple to specific courses. His research interests include the construction of environments for exploring specialized computational domains. Current projects include interactive environments in support of combinatorics, convex analysis, and the exact solutions of partial differential equations.

jsdevitt@daisy.uwaterloo.ca

EXPERIENCES WITH MAPLE IN ENGINEERING EDUCATION AT THE UNIVERSITY OF WATERLOO

Mustafa Fofana, Ian LeGrow, Stephen Carr
Engineering Education Research Centre, University of Waterloo, Waterloo ON, Canada

Abstract

At the University of Waterloo, several faculty members in Engineering have attempted to introduce Maple software into the engineering curriculum. Most of these attempts have been conducted with the assistance of the Engineering Education Research Centre. This paper summarizes some of the difficulties, successes, and lessons learned in these ventures, and provides some recommendations for future endeavours.

1. Introduction

The recent availability of symbolic computation languages and their demonstrated effectiveness in academic applications has fostered several enthusiastic attempts to introduce Computer Algebra Systems (CASs) into the engineering curriculum at the University of Waterloo. Because of its availability and the existence of a local support structure, Maple Software has been the package of choice at the University of Waterloo. Courses in which Maple software has been introduced are: Ordinary Differential Equations, Partial Differential Equations, Structural Mechanics, Dynamics, System Modelling, Linear Control Systems, etc. These courses are typically offered in the second or third years of the engineering curriculum.

The introduction of Maple software has been at the initiative of the instructors of these courses, and with the support of the facilities and expertise offered by the Engineering Education Research Centre (EERC).

The EERC has a mandate to support the effective use of new information technologies in engineering education. Engineering Computing and the EERC currently support networks of Unix workstations and Macintoshes in specialized teaching laboratories. In addition, a large network of PCs is available for general use by engineering undergraduates.

The motivations for introducing Maple into the engineering curriculum are many. Perhaps the primary reason is to remove the tedium associated with "hand" calculations, freeing the student to explore the subject material through a "What if .... ?" process which enhances understanding. Maple's two and three-dimensional plotting capabilities, though lacking at the time of the initial trials, was a motivating factor in subsequent efforts. An additional motivation for one of the earlier projects (Lee and Heppler [1990]) was to introduce "higher level" concepts into the curriculum, such as an appreciation for the difficulties of actual engineering design. Such attempts at introducing deliberately under-defined design problems would be untenable within the time constraints of an average undergraduate course without the powerful analytical capabilities of CASs and complimentary numerical analysis packages.

Mathematical Computation with Maple V:
Ideas and Applications
Tom Lee, Editor
©1993 Birkhäuser Boston

In this paper, we first describe the organization of courses which have incorporated Maple, including several examples from courses in differential equations (in which members of the EERC, (Austin and Fofana [1993]) had effectively participated in enhancing the teaching of differential equations). A summary of some of the problems encountered and recommendations for future ventures will be categorized.

2. Course Organization

In all courses, Maple software was introduced in optional tutorial sessions as a supplement to regular lectures, assignments and tutorials. Because this approach imposes an additional workload for the students, the use of Maple software was not mandatory. However, in the case where challenging design projects were assigned, this was significant incentive for the students to participate (Heppler [1989, 1990]).

The computing platform selected for all but one course was the Project Beacon Laboratory, a network of 15 Unix workstations. The remaining courses selected the Project Libra Laboratory, a network of 15 Macintosh II series workstations. The choice of computing platform was for the most part determined by the superior computing power of the Unix workstations compared to the other platforms, along with the lack of complimentary numerical analysis software the Libra Lab. Note, however, that within the last year hardware improvements to the PC network has also made it a viable platform option. Since the students were less familiar with the Unix operating system and supporting software such as text editors, it was further necessary to include familiarization tutorials on these subjects. The typical organization of the optional Maple sessions was:

- Introduce the operating system (approximately 1/2-1 hours);

- Introduce Maple (approximately 1/2-1 hours);

- Continued tutorial sessions—ranging from unstructured question-and-answer periods to structured, course-related examples (approximately 10 one hour sessions);

- Unstructured lab time (accessible 24 hours per day).

In the differential equations courses, examples were provided from a library compiled by the EERC teaching assistant (T.A.). Some of these examples worked out with the students are as follows:

(Q1). Define differential equation

```
> de1:=diff(y(x),x)-y*cot(x)=2*x*sin(x);
```

$$del := \frac{\partial}{\partial x}(y(x)) - y\cot(x) = 2x\sin(x)$$

General solution

```
> ygs1:=dsolve(de1,y(x));
```

$$del := y(x) = \sin(x)x^2 + \sin(x)C1$$

Graphs of y(x) for arbitrary values of C1

```
> s:= {};
```

$$s := \{\}$$

```
> for i from -4 by 2 to 30 do
> C1:= 0.5*i;
> s:=s union {sin(x)*x^2 + C1*sin(x)};
> od:
> plot(s,x = -Pi..2*Pi);
```

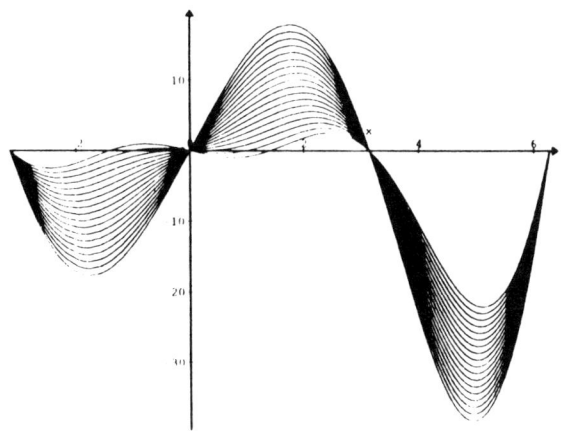

(Q2). Define differential equation

> de2:=diff(y(x),x$2)-4*diff(y(x),x) +
                    3*y(x)=5*cos(x);

$$de2 := (\frac{\partial^2}{\partial x^2} y(x)) - 4(\frac{\partial}{\partial x} y(x))$$
$$+ 3y(x) = 5\cos(x)$$

General solution

> ygs2:=dsolve(de2,y(x));

$$ygs2 := y(x) = \frac{1}{2}\cos(x) -$$
$$\sin(x) + C1 e^{3x} + C2 e^x$$

Checking the general solution

> subs(y(x)=rhs(ygs2),de2);

$$(\frac{\partial^2}{\partial x^2} \frac{1}{2}\cos(x) - \sin(x) +$$
$$C1 e^{3x} + C2 e^x) - 4(\frac{\partial}{\partial x} \frac{1}{2}\cos(x)$$
$$-\sin(x) + C1 e^{3x} + C2 e^x) +$$
$$\frac{3}{2}\cos(x) - 3\sin(x) +$$
$$3 C1 e^{3x} + 3 C2 e^x = 5\cos(x)$$

> simplify(");

$$5\cos(x) = 5\cos(x)$$

#(Q3). Define differential equation

>   de3:=x^2*diff(y(x),x$2)+5*x*
diff(y(x),x)+4*y(x)=x^2+16*(ln(x))^2;

$$de3 := x^2 (\frac{\partial^2}{\partial x^2} y(x)) +$$
$$5 x(\frac{\partial}{\partial x} y(x)) + 4 y(x) =$$
$$x^2 + 16\ln(x)^2$$

General solution

> ygs3:=dsolve(de3,y(x));

$$ygs3 := y(x) = \frac{1}{16 x^2} (x^4 +$$
$$64\ln(x)^2 x^2 - 128\ln(x) x^2 +$$
$$96 x^2 + 16 C1 + 16 C2 \ln(x))$$

> expand(");

$$y(x) = \frac{1}{16} x^2 + 4\ln(x)^2 -$$
$$8\ln(x) + 6 + \frac{C1}{x^2} + \frac{C2\ln(x)}{x^2}$$

Checking the general solution

> subs(y(x)=rhs(ygs3),de3);

$$x^2 (\frac{\partial^2}{\partial x^2} \frac{1}{16} \%1) + 5 x(\frac{\partial}{\partial x} \frac{1}{16} \%1)$$
$$+ \frac{1}{4} \%1 = x^2 + 16\ln(x)^2$$

$$\%1 := \frac{1}{x^2} (x^4 + 64\ln(x)^2 x^2 -$$
$$128\ln(x) x^2 + 96 x^2 +$$
$$16 C1 + 16 C2 \ln(x))$$

29

> simplify (");

$$x^2 + 16 \ln (x)^2 = x^2 + 16 \ln (x)^2$$

(Q4). Define differential equation

> de4:=diff(y(x),x$2)+4*diff(y(x),x)-
        5*y(x)=x^2+4*x;

$$de4 := (\frac{\partial^2}{\partial x^2} y(x)) + 4 (\frac{\partial}{\partial x} y(x))$$
$$- 5 y(x) = x^2 + 4 x$$

Define two initial conditions

> ic1:=y(0)=1;

$$ic1 := y(0) = 1$$

> ic2:=D(y)(0)=0;

$$ic2 := D(y)(0) = 0$$

Note that D is the differential operator

Particular solution

> yps4:=dsolve({de4,ic1,ic2},y(x));

$$yps4 := y(x) = -\frac{1}{5} x^2 - \frac{28}{25} x -$$
$$\frac{122}{125} + \frac{11}{6} e^x + \frac{107}{750} e^{-5x}$$

Checking the particular solution

> subs(y(x)=rhs(yps4),de4);

$$(\frac{\partial^2}{\partial x^2} - \frac{1}{5} x^2 - \frac{28}{25} x - \frac{122}{125} +$$
$$\frac{11}{6} e^x + \frac{107}{750} e^{-5x}) + 4 (\frac{\partial}{\partial x} -$$
$$\frac{1}{5} x^2 - \frac{28}{25} x - \frac{122}{125} + \frac{11}{6} e^x +$$
$$\frac{107}{750} e^{-5x}) + x^2 + \frac{28}{5} x + \frac{122}{25} -$$
$$\frac{55}{6} e^x - \frac{107}{150} e^{-5x} = x^2 + 4 x$$

> simplify(");

$$x^2 + 4 x = x^2 + 4 x$$

(Q5). Define differential equation

> de5:=diff(y(x),x$3)-2*diff(y(x),x$2)-
        5*diff(y(x),x)+6*y(x)=exp(3*x);

$$de5 := (\frac{\partial^3}{\partial x^3} y(x)) -$$
$$2 (\frac{\partial^2}{\partial x^2} y(x)) - 5 (\frac{\partial}{\partial x} y(x))$$
$$+ 6 y(x) = e^{3x}$$

Define three initial conditions

> ic1:=y(0)=0;

$$ic1 := y(0) = 0$$

> ic2:=D(y)(0)=0;

$$ic2 := D(y)(0) = 0$$

> ic3:=D(D(y))(0) = 1;

$$ic3 := D^{(2)}(y)(0) = 1$$

Particular solution

> yps5:=dsolve({de5,ic1,ic2,ic3},y(x));

$$yps5 := y(x) = \frac{3}{100} e^{3x} + \frac{1}{10} xe^{3x} - \frac{1}{12} e^x + \frac{4}{75} e^{-2x}$$

Checking the particular solution

> subs(y(x)=rhs(yps5),de5);

$$\left(\frac{\partial^3}{\partial x^3} \frac{3}{100} e^{3x} + \frac{1}{10} xe^{3x} - \frac{1}{12} e^x + \frac{4}{75} e^{-2x}\right) - 2\left(\frac{\partial^2}{\partial x^2} \frac{3}{100} e^{3x} + \frac{1}{10} xe^{3x} - \frac{1}{12} e^x + \frac{4}{75} e^{-2x}\right) -$$

$$5\%1 + \frac{9}{50} e^{3x} + \frac{3}{5} xe^{3x} - \frac{1}{2} e^x + \frac{8}{25} e^{-2x} = e^{3x}$$

$$\%1 := \frac{\partial}{\partial x} \frac{3}{100} e^{3x} + \frac{1}{10} xe^{3x} - \frac{1}{12} e^x + \frac{4}{75} e^{-2x}$$

> simplify(");

$$e^{3x} = e^{3x}$$

It is noteworthy that Maple software allowed solutions to assignment and tutorial problems to be distributed as "solution templates" written in the Maple language. This approach forced the students to actively work through the solutions themselves, simultaneously reinforcing Maple syntax and course concepts.

3. Problems Encountered

Incorporation of Maple software into the engineering curriculum has met with mixed success. A summary of the main difficulties follows:

● Students often became frustrated with having to learn a new operating system and text editor (i.e. Unix and vi-editor).

● The Maple Symbolic language has a relatively steep learning curve. Students found the Maple syntax to be "unnecessarily strict" (see Heppler 1990, SDE 353 course critique), referring to ":=" and the terminating ";". The language also includes unfamiliar data structures such as sets and lists as separate from arrays. Most other complaints have been addressed in subsequent versions of the program, with the notable exception of the inability to list user-assigned variable names.

● Students were daunted by the broad scope of Maple, often finding it difficult to find the proper functions necessary to complete an operation.

● There was little incentive to learn Maple and a new operating system. Students wanted to ensure that is was worth their while to put in the extra effort to do so.

● Students complained that too much new information was presented at once. In many courses, the introductions to Maple software and the Unix operating system were presented together in a single one or two-hour session which was overwhelming to the students. Also, this was done at the beginning of the semester when unfamiliar course material being presented in the classroom.

● In those course-experiments which were initiated by the EERC rather than by the course instructor, there was often a lack of continuity between the terminology and examples used in the classroom and those presented in the optional Maple sessions.

4. Recommendations

These recommendations are based on comments by instructors (professors), students' critiques, and personal experiences. These are:

(i) It is important not to oversaturate students with information. This happened when the students were introduced to a new computer system at the same time as Maple software. For the ODEs course the students were introduced to Unix and Maple in one session. Many of the students became confused with the Unix operating system. This confusion was exacerbated by introducing Maple, leaving many students frustrated. For future classes, a tutorial dedicated exclusively to teaching the operating system, its basic commands, and the use of a text editor is recommended. For the controls course only 27% of the students had used Unix previously; most students were familiar with DOS. Newer DOS machines are sufficiently powerful to operate Maple for Microsoft Windows, putting Maple in a familiar computing environment. Following the operating system tutorial, a subsequent tutorial would then be used to teach Maple. The controls course had a two hour session which the professor deemed too short.

(ii) Students' appreciation and desire to use Maple was higher in courses where they were already comfortable with the mathematical theory. In the ODEs course, where Maple was being used to help teach the theory, students who were having difficulty with the material found Maple to be an additional burden. In contrast, the students in the controls course found Maple extremely useful for (Lee and Heppler [1990]):

- Laplace and inverse Laplace transforms,

- Plotting,

- Factoring polynomials,

- Differentiation/Integration of rational functions,

- Expansion of partial fractions.

These were all concepts the students were familiar with prior to the course. The majority of the controls students said that Maple software did not teach them control theory; it simply allowed them to quickly confirm ideas about the theory learned in class or from the text. The successful application of Maple software to a course depends strongly on the course curriculum. Those courses in which the material is based on mathematics learned prior to the course are most successful.

(iii) Communication between the instructors of the courses and the T.A.s (teaching assistants) regarding tutorial content is essential. This ensures that the T.A.s use the same terminology as the instructors and present Maple examples that illustrate lecture material. Many students in some of the courses felt the examples they were shown in the Maple sessions were not close enough to the instructors' blackboard examples. Coordinating the Maple tutorials to lectures makes the use of Maple more relevant to the course and increases student participation in the tutorials.

(iv) Students must receive motivation from the course instructor to use Maple. If Maple is introduced as an option and the instructor does not require its use for any assignments or projects then students are not encouraged to surmount the initial learning curve. Motivation should consist of classroom encouragement and discussion of the use of Maple to solve sample problems. Some assignment problems should also be tailored to being solved using Maple. This can be done by using parameters in equations instead of numerical coefficients, encouraging students to experiment with and plot different solutions.

(v) Students expressed frustration over trying to find the commands to complete desired operations. The Maple software help facility contains detailed descriptions of every command along with examples. However, finding the desired command is an arduous task. To overcome this difficulty, a listing of the most common commands for a given course should be prepared for the students along with common tricks and tips.

References

Austin, H., and Fofana, M. S., " Maple as a Tool for Enhancing the Teaching of Ordinary Differential Equations to Second Year Civil Engineering Students," Proceedings of the Sixth Instructional Show & Tell for Ontario Universities and Colleges- Sharing Innovations for Teaching and Learning, University of Guelph, Guelph, 1993.

Lee, T., and Heppler, G. R., "Algebra Systems for Enhancing the Learning Environment for Control Systems," Proceedings of the ASEE Annual Conference, Education Research and Markets Division, Toronto, Ontario, Canada 1990.

Heppler, G. R., Course Critique of Systems Design (SD) 352 - Linear Control Systems, Internal Report, 1989.

Heppler, G. R., Course Critique of Systems Design (SD) 352 - Linear Control Systems, Internal Report, 1990.

Biographies

Mr. Stephen Carr is a doctoral student in the Department of Systems Design Engineering, Mr. Ian LeGrow is a fourth year Mechanical Engineering undergraduate student and Dr. Mustapha S. Fofana is a research associate in the Department of Civil Engineering, University of Waterloo. Mr. Carr's research interests include modelling and simulation of engineering systems using graph theory, with applications to design based on reliability and quality criteria. Mr. LeGrow studies flexible robot arms and control of nonlinear systems. Dr. Fofana's research interests are delay dynamical systems and their applications to machine-tool chatter; the use of Maple, Mathcad and Mathematica for enhancing the teaching of mathematics, mechanics and dynamic stability. All of the three authors work for the Engineering Education Research Centre (EERC) on a part time basis. The authors may be reached at

Engineering Education Research Centre
University of Waterloo, Waterloo,
Ontario N2L 3G1,

scarr@watserv1.uwaterloo.ca

msfofana@sail.uwaterloo.ca

iclegrow@sail.uwaterloo.ca

ON INTEGRATING COMPUTERS INTO THE PHYSICS CURRICULUM

Ronald L. Greene
Department of Physics, University of New Orleans, New Orleans LA, USA

Abstract

In this paper I present a case for using a unifying computational tool in core upper division physics lecture courses to help elucidate the physical principles discussed. I argue that a computer algebra/calculus system such as Maple is the most appropriate software class for such a role, and discuss several guidelines for its use. Some experiences acquired in using Maple V in junior level classical mechanics at the University of New Orleans are used to illustrate the argument. Three problems taken from the course are presented to illustrate the use of Maple.

Introduction

At most colleges and universities in the U.S. the lecture portion of the core undergraduate physics curriculum has remained substantially unchanged for over 20 years. In particular, we physics professors have been very slow to incorporate the use of computers into our lecture courses, despite the fact that we routinely use them in our research. Historically, technological limitations (expensive hardware and poor quality pedagogical software), together with faculty inertia, have discouraged us from breaking with the *status quo*.

Now, however, the technological barriers have all but disappeared. Hardware has become quite affordable, even for poorer departments. Excellent demonstration software and textbooks on applying computers to physics problems and simulations have appeared.[1-8] Things are

Mathematical Computation with Maple V:
Ideas and Applications
Tom Lee, Editor
©1993 Birkhäuser Boston

changing; the computer is entering our core lecture courses.[9] Despite this progress, however, Abraham, *et al.*[10] recently expressed a widely-held view when they wrote, "How to accomplish the integration of computing into the undergraduate [physics] major is an important question for the profession."

A Unifying Computational Tool

In my view, the most effective way of integrating computers into a diverse set of upper division lecture courses is to provide students with working knowledge of a unifying interactive computational tool. If students are taught to use the tool early in their studies, instructors of advanced classes can utilize this common computational background to present demonstrations or make computationally intensive homework assignments, without the necessity of spending class time teaching computer techniques rather than physics. Students could reap the pedagogical benefits of interactive computing without having to learn different software packages for each course. Furthermore, if the tool is powerful enough, students will find it helpful not only in their physics and other course work, but also in their future employment.

Computer Algebra/Calculus Systems (CACS) offer the most promise for providing a powerful, yet easy to use computational tool. (The potential for using CACS in physics courses has been recognized by a number of other physicists; see, for example, Refs. 11-15.) Their major advantage over other software tools such as spreadsheets and conventional programming languages or environments is that they are readily used in a non-declarative programming mode. This allows users to specify *what* is to be solved, without concern for the details of *how* it is to be solved.

Because of this, CACS are quite appropriate for a tool that is meant to aid students' understanding of physics, rather than programming. The emphasis shifts to analysis of the solution, such as verifying limiting cases and studying dependence upon the parameters of the problem.

So much of what we have traditionally taught in our core physics courses deals with analytical solutions. The symbolic abilities of CACS will allow us to incorporate their use into our courses with minimal required changes in the way the courses are now taught, which is necessary for acceptance by the majority of current physics faculty. However, their numerical and graphical abilities, and their interactivity will allow us to go beyond the bounds of conventional lecturing and the idealistic, analytically solvable problems to which we now limit ourselves.

CACS are very large and powerful programs. Many hours of experience are required to become proficient with them. Fortunately, only a small subset of the predefined functions need be learned by the student (teacher) to usefully apply a given CACS to learning (teaching) the material in undergraduate physics courses. No programming is required, although it can be used to extend the power of the system. To illustrate, I will shortly present some aspects of using Maple V in the junior level classical mechanics course at the University of New Orleans. I will list the subset of functions used and give examples taken from the course. First, however, it is important to discuss briefly a general approach toward using these programs in physics courses.

Philosophy of Use

My guiding principle for using a CACS, or any other computational tool, in an undergraduate course is this: **The primary purpose of the computational tool is to enhance students' understanding of physics.** This principle has led me to adopt several guidelines for incorporating the use of CACS in physics courses.

1) The effort required to learn to use the system should be minimal. This can be achieved by teaching students to use a highly restricted subset of the program. For example, a small subset of Maple V

is sufficiently powerful both to save tedious, time-consuming calculations in junior level classical mechanics and to allow instructors and students to take advantage of interactive computing to probe more deeply into the physics of a given problem.

2) Avoid trying to teach the typical physics major how to write programs with the language included in the system. More advanced students may want to get into programming, or instructors may want to introduce programming in specialized courses, but time spent programming in core undergraduate physics courses is for the most part time taken away from learning physics.

3) Emphasize the importance of checking solutions. All CACS still have bugs in them, some of which are serious. This is not a reason to avoid their use, since both students and teachers also make errors. However, there seems to be a tendency for students to accept results that come from a computer as correct, without the need to check them. Results can be checked using the CACS (since typically a different portion of the program code is used for checking a solution than for obtaining the solution), and/or by hand (since it is often easier to verify a solution by plugging it into the equation than to solve the equation in the first place). Beyond mathematical checking, however, students should be taught to examine their answers for physical plausibility, and to verify limiting cases.

4) Do not insist on using the CACS for things that are more easily done by hand. One of the things that can take the most time and require the most experience for success is cajoling the computer to yield the "simplest" form for an expression. In learning to use a CACS it can be very frustrating for a user to see an obvious simplification that the computer refuses to make. If simplification is necessary (It is not for such tasks as plotting or for intermediate results that will be used to get limiting cases.), and it is easier for students to work out the simplified form

on paper or in their heads, let them do so. The simplified form can be re-entered into the CACS if desired.

5) In so far as possible, the basics of the CACS should be taught using physics that is already familiar to the students, for obvious reasons. For example, early in the first semester of the typical classical mechanics sequence is a unit reviewing Newtonian physics of a particle. Most of the CACS functions that will be used in the complete classical mechanics course can be introduced here and used to solve and analyze the kinds of problems that students have already seen in their introductory course.

Probably the matter that will generate the most controversy with regard to using a CACS in physics courses is the question of how much mathematics students should be required to do by hand. The same type of question was debated years ago with regard to students' use of calculators. (Apparently this kind of debate goes back much further. A Mathematics professor friend of mine tells me that there was an outcry against students' use of chalk and slate in learning mathematics because of the deleterious effect it would have on their ability to do arithmetic in their heads.) Obviously, each teacher will decide for him/herself how much they will allow their students to rely on the computer for solving equations. My personal expectation is that within ten years virtually all solutions at the undergraduate level will be done by computer, and we instructors will be grading students on their understanding of physics as reflected in their ability to obtain the proper equations to solve, and their analysis of the results, rather than their mathematical manipulations.

Maple V in Classical Mechanics

Undergraduate classical mechanics at the University of New Orleans is taught as a series of two courses, one at the junior level and the other at the senior level. I taught the junior level course in the Fall 1992 semester, incorporating the Maple V system into the course.

The first few weeks of the course were conducted in the usual manner, covering introductory material on vectors and kinematics, and reviewing Newton's equations of motion through in-class and homework problems. I then proceeded to introduce the students to Maple and had them use it to solve many of the same problems that they had already done by hand.

Like most CACS, Maple V has a large number of predefined functions (2500 in the latest release). Fortunately, very few are needed to use the program productively in a classical mechanics course. For their basic Maple text and reference, students were required to purchase *First Leaves: A Tutorial Introduction to Maple V* by Char *et al.*[17] I spent two class periods introducing Maple V, discussing entering expressions, editing, basic operations, and most of the functions that would be used in the course. This corresponds to most of Chapter 1 and a third of Chapter 2 in the Maple

---

Table I. Basic Maple Functions

diff -- ordinary and partial differentiation
int -- indefinite and definite integration
solve -- solution of algebraic equations
dsolve -- solution of differential equations
subs -- substituting expressions into equations
eval -- forcing evaluation of an expression
evalc -- separating an expression into real and imaginary parts
evalf -- numeric evaluation of expressions
assign -- assign values to variables of a solution set
simplify -- generic simplification of expressions
limit -- verifying limiting cases
series -- small parameter expansions
unapply -- creating functions from expressions
plot -- plotting results

---

tutorial (about 65 pages total). Table I gives a listing of most of the basic functions that I recommended students use for their assignments. The brief explanation describes what the given function was used for in the course; many have other uses as well. In addition to these basic functions, there are a number of more specialized functions for manipulating expressions (expand, factor, combine, normal, convert, collect) or extracting parts of solutions (rhs, lhs, coeff, numer, denom). Students used some of these more specialized functions after they became comfortable with the basic Maple operations and functions.

The version of Maple V, Release 1 used in the course was installed on our departmental IBM RS/6000 workstation network, running AIX with Xwindows. However, every function we used is also available on the MS-DOS and Macintosh Student Versions, which are available from Brooks/Cole Publishing.

One characteristic of Maple is that if you type in a syntactically correct command with a function name which has not been defined (perhaps a misspelled or improperly cased name for one of the predefined functions), the program will return the expression as typed, with no comment. Every student was baffled by this at one time or another. However, they learned to carefully check their typing and function names when the expected answer failed to appear. More troublesome were the occasions when Maple was given a command that it apparently could not compute, and it simply returned a prompt for the next expression. This lack of explanation was annoying and puzzling to both students as well as the instructor.

Example Problems

In order to provide a clearer picture of how Maple can be used in the first classical mechanics course, I present three example problems below. The purpose of these examples is to give a flavor of how solutions can be obtained and analyzed using Maple. The first example is a homework problem taken from the review chapter of the mechanics textbook. The second is a problem taken from one of the exams, while the third is a look at a sinusoidally driven harmonic oscillator (which was presented in a class lecture).

Example 1

The system shown in Fig. 1 consists of two masses (M_1 and M_2) sliding on horizontal surfaces, each with a coefficient of friction μ. The third mass (M_3) is connected to a frictionless and massless pulley which is supported by a massless rope whose ends are tied to M_1 and M_2. The problem is to use Newton's 2nd law to find the tension in the rope.

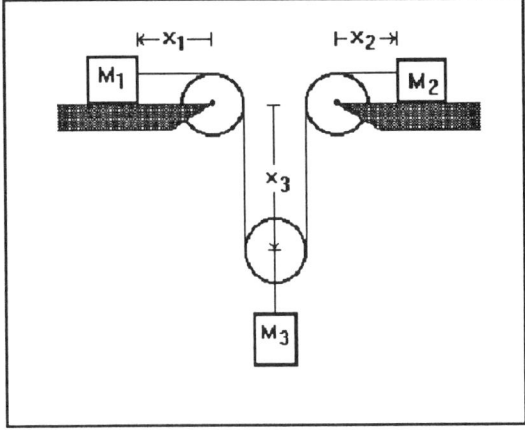

Figure 1. Sketch of the system of Example 1. Masses M_1 and M_2 slide toward each other with coefficients of friction μ.

Let us consider the case in which M_1 and M_2 are each initially moving toward the center of the figure. Applying Newton's 2nd law to the three masses (assuming positive directions given by the arrows in the figure) we get the following three equations:

$$\mu M_1 g - T = M_1 a_1$$
$$\mu M_2 g - T = M_2 a_2$$
$$M_3 g - 2T = M_3 a_3 \ .$$

There is, in addition, a constraint due to the fact that the length of the rope is constant. This requires that $x_1 + x_2 + 2 x_3 = $ constant, or

$$a_1 + a_2 + 2a_3 = 0 \ .$$

This example is typical of the kind of Newton's 2nd law problem that students work in their introductory course. Since all the forces and accelerations are constant, obtaining the solution requires simultaneously solving four linear algebraic equations, a trivial job with Maple. As instructors of the junior level classical mechanics course, we would like to insure that our students can correctly set up the coordinates and second law equations for these kinds of problems, and properly visualize limiting cases of the motion. However, the tedium of solving multiple equations by hand normally prevents us from assigning very many of these problems, and the likelihood of error in getting the solution diminishes the value of the exercise of examining limiting cases. Maple not only removes the tedium and reduces the chance of error, but it can aid in the examination of the analysis of the solutions.

Example 2

For the second example let us consider the motion of a projectile subject to a drag force proportional to the velocity. The problem is to calculate the position and velocity as a function of time, and to examine the effect of drag on the trajectory. Assume x and y are the horizontal and vertical axes with positive directions being to the right and upward, respectively. For this problem, Newton's 2nd law can be used to obtain an equation of motion which is a first order differential equation for $\mathbf{v}(t)$, rather than the more usual second order equation for $\mathbf{r}(t)$:

$$-mg\hat{\mathbf{y}}-b\mathbf{v}=m\frac{d\mathbf{v}}{dt}\ .$$

With some rearrangement this vector equation can be broken into the following x and y equations:

$$m\frac{dv_x(t)}{dt}+bv_x(t)=0$$

and

$$m\frac{dv_y(t)}{dt}+bv_y(t)+mg=0\ .$$

The initial values of velocity are taken to be $v_x(0)=v_{0x}$ and $v_y(0)=v_{0y}$.

To solve these equations with Maple, we first use the 'diff' function to enter the equations,

eqx := m*diff(vx(t),t) + b*vx(t) = 0;
eqy := m*diff(vy(t),t)+b*vy(t)+m*g=0; ,

and solve them with the 'dsolve' function. (At this point in the course I introduced the concept of scaling equations to show students how to reduce the number of parameters. However, I will not use scaling in this paper.) The solutions are found with the statement

sol := dsolve({eqx,eqy,vx(0)=v0x,vy(0)=v0y},
{vx(t),vy(t)}); .

(The equations are uncoupled, and thus can be solved separately if desired.) They can be checked with

simplify(eval(subs(sol,{eqx,eqy}))); ,

which results in

$$\{0 = 0\}\ .$$

If we assign the appropriate values to vx(t) and vy(t), we can verify the initial conditions:

assign(sol);
limit(vx(t),t=0);
limit(vy(t),t=0);

As always, it is important to verify limiting cases for which we know the answers. One obvious limiting case to check is that of no air resistance. This case can be obtained with

limit(vx(t),b=0);
limit(vy(t),b=0); ,

which yield the expected results.

Another useful limit to verify is the long-time limit. Physics tells us that as $t\rightarrow\infty$, $v_x(t)\rightarrow 0$, and $v_y(t)\rightarrow -mg/b$, the so-called terminal velocity. We were not able to verify this limit with Maple because I did not know how to specify that b/m is positive. Release 2 has solved this problem.

In order to graphically compare velocity and position as functions of time for different drag coefficients, it is convenient to create functions of b and t out of the expresssions which are the values of the Maple variables vx(t) and vy(t). This can be done with the 'unapply' function:

VX := unapply(vx(t),b,t);
VY := unapply(vy(t),b,t); .

[Capital letters have been used to avoid confusion with the Maple variables vx(t) and vy(t).] Given the velocity components as functions of time, we can then find the position components by integrating in the usual way. Assuming that the origin of the coordinate system is chosen at the initial position of the projectile, we define

X := (b,t) -> int(VX(b,s),s=0..t);
Y := (b,t) -> int(VY(b,s),s=0..t); .

Once we specify the parameters of the problem, the definitions for VX(b,t), VY(b,t), X(b,t) and Y(b,t) will allow us to plot the x- and y-components of position and velocity as functions of time or drag coefficient. For example, suppose we look at the special case of dropping a rock vertically from the edge of a cliff. We make the assignments (in SI units)

$$m := 1; \quad g := 9.8; \quad v0x := 0; \quad v0y := 0; \quad .$$

Maple will automatically incorporate these values for the parameters into our definitions for the position and velocity functions, allowing us to plot the results with little effort. (For some reason I do not yet understand, the values for vx(t) and vy(t) do not incorporate the new values of the parameters.)

We can directly compare the y-component of the velocity as a function of time for three different drag coefficients: b=0, 1, 10 kg/s. The command is

$$\text{plot}(\{\text{limit}(VY(b,t),b=0),\ VY(1,t),$$
$$VY(10,t)\}, t=0..1);\quad .$$

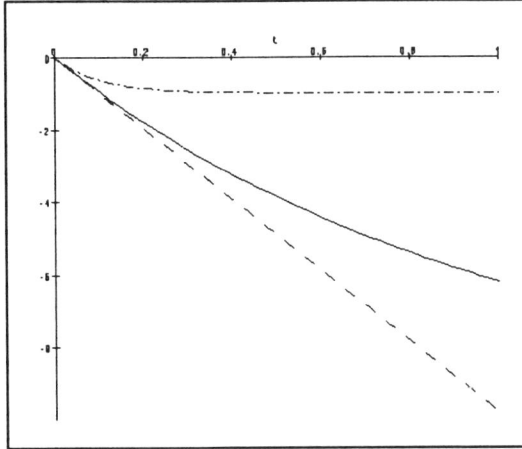

Figure 2. Vertical component of the velocity as a function of time for the projectile described in Example 2.

The results are shown in Fig. 2. The b=0 curve shows the expected linear dependence of the velocity upon time. The b=10 kg/s curve, which corresponds to a small density object like a styrofoam ball, clearly shows the rapid onset of terminal velocity.

To see the effect of drag upon the trajectory of a projectile in two-dimensions, we change the values of v_{0x} and v_{0y}, and use Maple's parametric plotting facility to plot Y(b,t) vs. X(b,t) over time, for several values of b. For example, the Maple sequence

$$v0x := 10; \quad v0y := 10;$$
$$\text{plot}(\{[\text{limit}(X(b,t),b=0),$$
$$\text{limit}(Y(b,t),b=0),t=0..2],$$
$$[X(.1,t),Y(.1,t),t=0..2],$$
$$[X(1,t),Y(1,t),t=0..2]\});$$

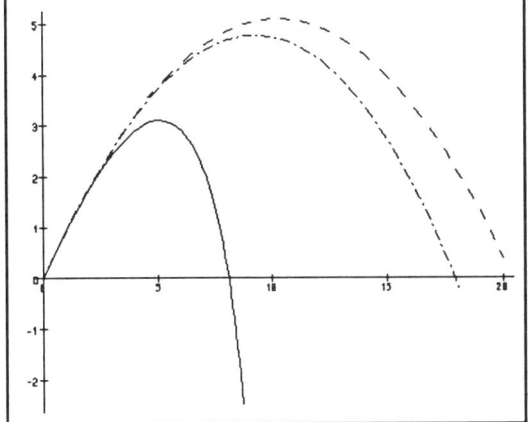

Figure 3. Trajectory of the projectile described in Example 2 for three different values of the drag coefficient.

yields the three curves shown in Fig. 3. The decrease in range and the increasing deviation from parabolicity with increasing b are quite evident in the figure.

Example 3

The last example problem is that of a sinusoidally-driven, underdamped harmonic oscillator. A lot of physics can be extracted from the harmonic oscillator in its many forms. However, for the sake of brevity, let us consider one specific aspect of the problem --- that of energy absorption by the oscillator from the driving force.

The following is the equation of motion for a damped, sinusoidally-driven oscillator (not the most general form):

$$F_o \sin(\omega t) - b\frac{dx(t)}{dt} - kx(t) = m\frac{d^2x(t)}{dt^2}$$

Dividing by m and rearranging terms yields

$$\frac{d^2x(t)}{dt^2} + 2\gamma \frac{dx(t)}{dt} + \omega_o^2 x(t) - \frac{F_o}{m}\sin(\omega t) =$$

where $\omega_o^2 = k/m$ and $\gamma = b/(2m)$. If we take conditions such that the initial position and velocity are zero, the following lines enter the equation into Maple, solve it, check the solution (including initial conditions), and assign the values for the position and velocity to the variables x(t) and v:

```
Dforce := F*sin(w*t);
eq := diff(x(t),t,t) + 2*gam*diff(x(t),t)
        + w0^2*x(t) - Dforce/m = 0;
sol := dsolve({eq,x(0)=0,D(x)(0)=0},x(t));
    simplify(eval(subs(sol,eq)));
            assign(sol);
            limit(x(t),t=0);
            v := diff(x(t),t);
            limit(v,t=0);
```

(The class of the solution --- underdamped, overdamped, or critically damped --- depends on the value of $\omega_o^2 - \gamma^2$. I had not found a way to specify the signs or relative magnitudes of parameters or combinations of parameters in Maple at the time that I did this in class. With the advent of Release 2, the underdamped case can be specified by including statements

```
assume(gam>0);
additionally(gam^2-w0^2<0);   .)
```

We could examine the solution at this point and discuss transient and steady state terms. However, we will instead concentrate on the energy of the oscillator. We first define the kinetic and potential energies:

```
K := m*v^2/2;
U := m*(w0*x(t))^2/2;
```

In order to plot results, we need to assign numerical values to the parameters; for example,

```
m:=1; w0:=1; F:=1; gam:=1/10;
```

Note that this corresponds to a lightly damped oscillator. Furthermore, by taking γ to be 1/10 rather than 0.1, we avoid problems arising from rounding of floating point numbers (which can lead to small, but non-zero imaginary parts).

Let us first look at the total energy of the oscillator as a function of time for three values of

ω. This is easily done by first defining the total energy to be a function of ω and t:

```
En := unapply(evalc(K+U),w,t);   .
```

The purpose of the 'evalc' function in this definition is to force Maple to separate the real and imaginary parts of the K+U sum so that that the total energy is recognized as real. Note that it is only after the assignment of ω_o and γ that Maple can unambiguously perform the evalc. (If the result contained a small imaginary part to due rounding of floating point numbers, we would also have to extract the real part to perform the plot.)

If we now plot the total energy as a function of time for $\omega=.5$, .9, and 1 rad/s (The last value corresponds to resonance.),

```
plot({En(.5,t), En(.9,t), E(1,t)},
        t=0..10/gam,numpoints=200);
```

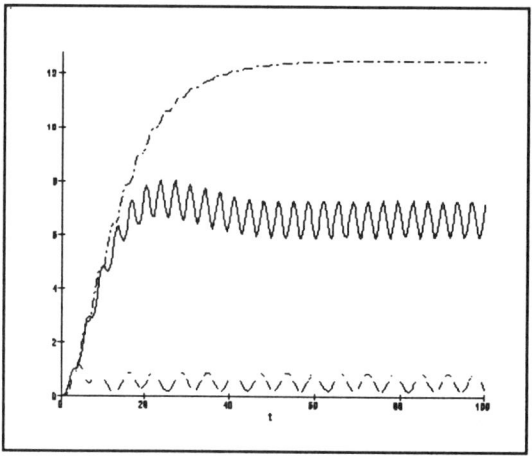

Figure 4. The energy of the sinusoidally-driven, underdamped harmonic oscillator of Example 3 as a function of time, for three values of the driving frequency.

we get the result shown in Fig. 4. At this point in the course the students had already seen that the energy of the undriven, undamped oscillator is constant, and that of the undriven, damped oscillator decreases in time. Thus it was clear to them that the initial increase in energy in this case comes from the driving force. The meaning of steady state is also clarified by the long-time behavior of these curves.

The effects of resonance, which are suggested

in Fig. 4 by the fact that energy absorption at $\omega = 1$ rad/s is much larger than the other two values plotted, can be illustrated more clearly by examining the power absorbed by the oscillator from the driving force, averaged over one period. It is instructive to compare the power absorbed for different values of damping, so we will unassign γ and define the average power absorbed as a function of γ. We treat two cases with different starting times below (By $t = 10/\gamma$ the transients have effectively died out, as Fig. 4 showed, whereas at $t = 0$ they are very much present.):

```
gam: = 'gam';
T : = 2*Pi/w;
int(Dforce*v,t=t0..t0+T)/T;
AvgPower : = unapply(simplify("),t0,gam,w);
P0 : = (gam,w) -> Real(AvgPower(0,gam,w));
Pss : = (gam,w) ->
    Real(AvgPower(10/gam,gam,w,t)); .
```

If now the average power is plotted as a function of ω, the effect of a driving force whose frequency is at or near resonance can be clearly seen.

```
plot({P0(1/10,w),Pss(1/10,w)},w=0..4); .
```

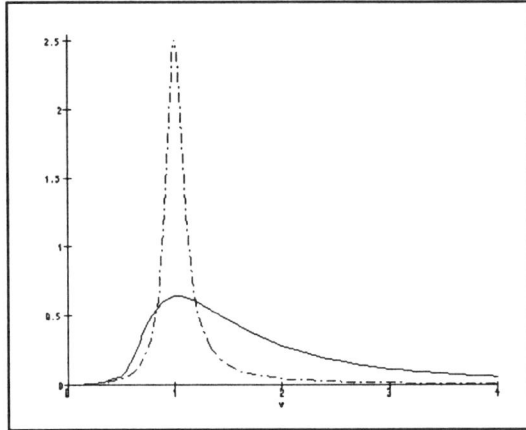

Figure 5. Average power absorption of the oscillator in Example 3 as a function of the driving frequency. The two curves represent the average over periods starting at $t = 0$ and $t = 10/\gamma$ (near-steady-state).

The result is shown in Fig. 5, for the case $\gamma = .1$, which corresponds to weak damping. Most mechanics texts show an average power absorbed vs. driving frequency at steady state, but not at earlier times when transients are still present. It is fairly easy to reproduce the steady-state-only case with Maple, should we desire. However, the Pss function is a good approximation to it.

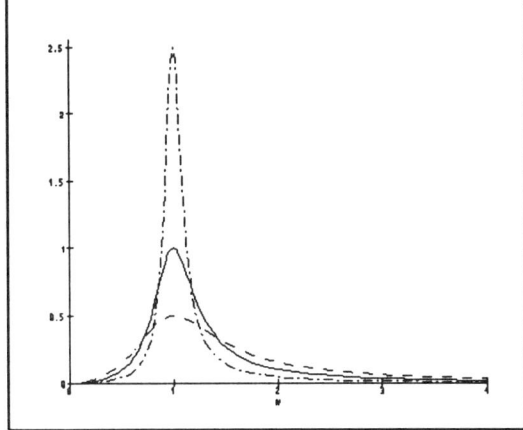

Figure 6. Average near-steady-state power absorption of the oscillator in Example 3 as a function of driving frequency. Three different values of damping are shown.

Finally, we show in Fig. 6 a comparison of near-steady-state power absorption for three different values of γ. The plot is that obtained from

```
plot({Pss(1/10,w),Pss(1/4,w),
    Pss(1/2,w)},w=0..4); .
```

The most obvious features, of course, are the strong power absorption near resonance ($\omega = \omega_0 = 1$ rad/s), and the large increase in resonant absorption as damping is decreased.

Additional Comments

Much of what we teach in our upper division physics courses is driven by the difficulty in obtaining results mathematically. The absence of analytical solutions can cause us to spend little time with a given problem, or to ignore it entirely. Even when analytic solutions exist they may be so tedious to work with that we similarly avoid the problem. The driven harmonic oscillator provides us an example of this latter type of problem. The mathematics required to study the energy of the oscillator and power absorption in the general case is simple, but extremely tedious. Its very detail tends to obscure the physics. This has caused us to omit interesting parts of the physics in our class discussions. However, as seen in the previous section, a CACS analysis allows us to look (for example) just as easily at energy of the oscillator and average power absorption when the transients are still present as when it has reached steady state. A similar argument can be made for using the numerical abilities of a CACS to study the physics of problems which have no analytical solutions. These software packages open new territory for us to explore in our classes without the necessity of spending hours filling blackboards with additional derivations or looking only at limited approximations. In addition, using a CACS at the level that I advocate for our undergraduate courses is sufficiently easy that we can assign such explorations to students with the confidence that they can learn interesting physics from the exercise.

References

1. Denis Donnelly, "CIP's First Annual Software Contest: The Winners," Computers in Physics **4**, 540-548 (1990).
2. Denis Donnelly, "CIP Announces Winners of Second Annual Software Contest," Computers in Physics **5**, 636-642 (1991).
3. Denis Donnelly, "CIP's Third Annual Software Contest: The Winners," Computers in Physics **6**, 686-691 (1992).
4. Steven E. Koonin, *Computational Physics* (Benjamin Cummings, Menlo Park, CA, 1986).
5. Harvey Gould and Jan Tobochnik, *An Introduction to Computer Simulation Methods: Applications to Physical Systems, Parts 1 and 2* (Addison-Wesley, Reading, MA, 1988).
6. Marvin L. De Jong, *Introduction to Computational Physics* (Addison-Wesley, Reading, MA, 1991).
7. Denis Donnelly, *MathCAD® for Introductory Physics* (Addison-Wesley, Reading, MA, 1992).
8. Charles Misner and Patrick Cooney, *Spreadsheet Physics* (Addison-Wesley, Reading, MA, 1991).
9. Jack M. Wilson, "Computer Software Has Begun to Change Physics Education," Computers in Physics **5**, 580-581 (1991).
10. Neal B. Abraham, James B. Gerhart, Russell K. Hobbie, Lillian C. McDermott, Robert H. Romer, and Bruce R. Thomas, "The undergraduate physics major," Am. J. Phys. **59**, 106-111 (1991).
11. Richard E. Crandall, "Symbolic Software Supports New Adventures in Physics," Computers in Physics **5**, 576-579 (1991).
12. Patrick Tam, "Physics and Mathematica™", Computers in Physics **5**, 342-348 and 438-442 (1991).
13. Cyrus Taylor, "*Mathematica* in the Classroom," Computers in Physics **5**, 16-21 (1991).
14. David M. Cook, Russell Dubisch, Glenn Sowell, Patrick Tam, and Denis Donnelly, "A Comparison of Several Symbol-Manipulating Programs: Part I," Computers in Physics **6**, 411-420 (1992).
15. David M. Cook, Russell Dubisch, Glenn Sowell, Patrick Tam, and Denis Donnelly, "A Comparison of Several Symbol-Manipulating Programs: Part II," Computers in Physics **6**, 530-540 (1992).
16. Bruce W. Char, Keith O. Geddes, Gaston H. Gonnet, Benton L. Leong, Michael B. Monagan, Stephen M. Watt, *First Leaves: A Tutorial Introduction to Maple V* (Springer-Verlag, New York, NY, 1992).

Ronald L. Greene is a University of New Orleans Research Professor in the Department of Physics. He has been actively involved in theoretical and computational studies in plasma spectroscopy and semiconductor physics at various times in his career. Most recently his research interests have been in the area of neural networks and associative memory. He is leading the effort at the UNO Department of Physics to incorporate computer algebra and calculus into the undergraduate physics curriculum.

USING MAPLE AND THE CALCULUS REFORM MATERIAL IN THE CALCULUS SEQUENCE

David C. Royster
Department of Mathematics, University of North Carolina, Charlotte NC, USA

Abstract

Computer Mathematics Systems and Calculus Reform Material are topics of discussion in many Mathematics Departments. There are many questions about how to best use each of these in mathematics courses. Research into the way that college students learn mathematics will have a great impact on the way that we teach mathematics and how we use these tools. We discuss some of the questions and offer my experience in using both tools together to have one reinforce the other.

Introduction

In 1986 a call went forth to consider seriously the Calculus sequence. In response to that call a conference was held at Tulane University and issuing forth therefrom came *Toward a Lean and Lively Calculus* [Doug86]. This was the first concerted effort to look at the calculus sequence and see what, if anything, needed to be changed. Many threads were started at this workshop that are today weaving their way through the mathematics curriculum.

The National Science Foundation listened to the call made in this workshop and seeded projects to make changes in the Calculus curriculum. In the intervening 7 years major changes to the Calculus curriculum have been proposed. While not all of the projects started with the same goals, certain ideas came to the forefront in each of the projects. One of the basic tenets that arose in each project was that each of the basic concepts of Calculus should, if at all possible, be presented symbolically, graphically, and numerically. This idea of multiple representations is not new to Mathematics or Mathematics Education, but it was not used to any great extent in the Calculus

Mathematical Computation with Maple V:
Ideas and Applications
Tom Lee, Editor

sequence. The derivative was developed graphically as the slope of the tangent line and the tangent line as the limit of secant lines, but when it came time to work with the derivative we would retreat to our symbolic representation. The integral is introduced as area under a curve—adding up the areas of small rectangles based on the value of the function at some point in each subinterval: the Riemann sum. We would do a Riemann sum for very few functions and immediately would retreat to the Fundamental Theorem of Calculus as a tool for evaluating definite integrals. In fact most students did not see the mathematical excitement of the Fundamental Theorem of Calculus. They saw it only as a tool for evaluating integrals.

Using multiple representations to a better purpose was seen as one way in which the Calculus curriculum should change—to give equal importance to each of the different representations and work with the different representations throughout the course. However, numerical representations took a long time to compute and were rarely done. To find a zero of a function we would use Newton's method if it converged quickly, but would rarely use the bisection method, though pedagogically simpler to motivate. The time required to do these computations was a restricting factor. In addition, it was difficult to use the graphical representation, because it took a lot of time to draw the graphs and then they were not necessarily drawn well.

At this time there were computer programs available that could deal with the graphical representation, the numerical representation, and the symbolic representation. They were not widely available and the delivery of these programs to a large audience was a concern. At about this time the graphing calculator fortuitously arrived on the scene. Also in the late 1980s, the computer mathematics systems began to expand their different abilities and availability on different platforms and their user in

43

terfaces. At the Tulane Conference workshop Donald Small and John Hosack [SH86] looked at the possibility of using a Computer Mathematics System in the Calculus sequence. They raised a number of questions and these questions have been discussed extensively in [SH86, Dev90, Harv90, Ral90, Zorn90], and in many of the articles in [Kar92]. How should the computer be used in the Calculus sequence? When should it be used in the Calculus sequence? Should it be used in the Calculus sequence? In [HLLL92] the authors mention that computer mathematics systems "were not initially designed or perceived as pedagogical tools but rather as professional tools." Like most of the technologies introduced, their influence spilled into the academic arena. At this point we had to admit their existence, though we did not know how to use them wisely or too well.

Here, following the suggestion of Dick Shumway, we call such a system a *Computer Mathematics System*(CMS) rather than a *Computer Algebra System* because it does so much more than algebra. Among computer mathematics systems, in addition to *Maple*, are programs such as *Axiom*, *Derive*, *Macsyma*, and *Mathematica*. These are mentioned because they are commercially available for a variety of platforms[1]. Their latest incarnations allow the student to do symbolic manipulation at a basic level of algebra, but more importantly, at higher levels. They handle graphical output much better and have the ability to give high precision numerical output. They have been classified as super-calculators, but should be considered as much more.

Several NSF funded projects—Calculus Consortium at Harvard, St. Olaf Calculus Project, Oregon State Calculus Project, Purdue Calculus Project, Calculus in Context Project(Five Colleges, MA), Project CALC(Duke University)—have decided to use multiple representations for each of the concepts in presenting the Calculus. This development of the Calculus reform material paralleled the development of the computer mathematics systems and their use in the Calculus classroom. The concerns raised by the use of the CMS in the classrooms were being addressed in the reform material, though of course not

independently. Since the arrival of the CMS coincided with this era of change in calculus teaching, we have embraced it as a means to bring about change both in the curriculum and, especially, in the way that we teach and the way that the student learns mathematics.

This multiple-representation approach to Calculus is well-suited for the use of *Maple*. *Maple* very easily allows us to use each of these representations in studying functions, limits, derivatives, integrals, and other topics in the reformed Calculus curriculum. *Maple* allows the students to tackle more realistic, more complex, and more interesting problems.

Questions remain about the appropriate use of *Maple*.

- When should *Maple*, or any computer mathematics system, be introduced to the students in the Calculus sequence?
- How much symbolic manipulation is appropriate for the first semester of Calculus?
- How much symbolic manipulation is appropriate for the second semester?
- Is *Maple* inappropriate for the Calculus sequence?
- What does the research say about the use of computer mathematics systems with respect to understanding of the concepts from Calculus?
- Can the use of *Maple*, in conjunction with the Calculus Reform material, help students understand the concepts more easily?

These questions are addressed to the Calculus course. We are not certain how to begin to answer these types of questions about the learning processes of students at lower levels. There are many things that we do not know about the process of learning algebra and its role in the learning of Mathematics. We are only beginning to make progress at the Middle School level and the High School level in learning how students construct their own mathematical systems. The universities are not dealing with students at this cognitive level, but are dealing with students who have constructed their algebra and their mathematical systems. Newer research by Rachlin in the Hawaii Algebra Project and by Heid, Sheets, and Matras [HSM90] is giving us a better understanding of the process that students use to construct their mathematics and their *algebra*. We are coming to

---

[1] with the exception of *Derive* available only on DOS based machines, but such a wide variety of DOS based machines.

the realization that their *algebra* may not be *our algebra*.

- Is there some way in which the computer mathematics systems can direct the students' learning and the construction of the students' mathematics?
- Is it possible that having students use a computer mathematics system will retard their natural learning of the pattern recognition that is inherent in the learning of algebra?
- Are the algebraic abilities and the pattern recognition gained in learning these abilities necessary in later mathematics?
- Will these abilities not be present at a later date when they are required to understand deeper mathematical concepts?

We can spend many hours, days, and weeks arguing about the appropriateness of using technology in the classroom, let me quote a former engineer who responded to a post to the GRAPH-TI LISTSERV on the Internet:

> *After a career of 20 years in Research, Development, Testing and Evaluation in the Department of Defense, I find it appalling that technology would not be allowed in engineering courses.*
>
> *In my experience an engineer who is not fully aware of technology and how to use it is not a viable candidate for a job. When I hired an entry level engineer, he was expected to be proficient in the available technology. Time and dollars do not permit the use of long division and reliance on paper and pencil memory. They require the use of the best technology available.*
>
> *In all those years, I never met a real-world engineering problem that was solvable using only paper and pencil. The ability to use technology is required and demanded.*
>
> *Our students are required to work and live in a world of technology, not one populated by old Greeks drawing figures in the sand. Newton wrote, "If I have seen further it is by standing on the shoulders of giants." The small black, blue or gray boxes that are today's technology are the epaulets on the shoulders of the giants upon which our students will stand to see further and perhaps to go where no one has gone before.*

The State of the Computer Mathematics System

Does a computer mathematics system make a difference in a student's learning mathematics? Not a great deal of research has been published to date. A number of 1990–1992 dissertations address the area of the use of technology in the collegiate mathematics setting. Of those, a small number deal with Calculus and computer mathematics systems [HLLL92, Heid88, Palm91, Jud90]. In these experiments the computer mathematics system was used in different ways. The researchers were not always able to offer the computer mathematics system to the student in a same format. Nonetheless, the findings of these studies indicate that the students in the groups using the computer mathematics system learn the concepts better than the control groups and do as well, or better, than the control on standard paper-and-pencil tests. Charlene Beckmann [Beck90] studied the impact of a graphical interface with the students in different calculus courses (N=163). She found that developing the calculus concepts through the use of a graphical representation system (specifically computer graphics) positively affected student understanding and interest without necessarily negatively influencing skill acquisition.

These results are in line with similar studies on the connection between the use of the scientific and graphing calculator in learning mathematics and the numerical abilities of students at all levels. For those students who took a class and learned their mathematics using the calculator, the calculator had a neutral impact or a slightly positive impact on their ability to perform, unaided by the calculator, on standard paper-and-pencil tests.

The State of the Calculus Reform Material

Does the Calculus Reform material make a difference in how students learn calculus and what they learn? Again, there is anecdotal evidence that it does have a positive impact on those students who complete the course, regardless of the technology chosen for the course, *cf.* [Hart92]. There have not been, to date, any large studies published on these effects. Studies are underway in the Purdue Project, Project CALC, and at UNC Charlotte. Other studies are in progress.

Part of the difficulty in this area of study is to determine a good baseline for the mathematical knowledge of the college student. In addition we do no know how college students learn mathematics? Dubinsky [Dub92] outlines four possible ways in which students learn mathematics:

- **Spontaneously:** Students learn mathematics individually and spontaneously.
- **Inductively:** Students learn inductively by doing many examples, extracting common features, and organizing this information in their minds.
- **Constructively:** Students learn by making mental constructions to deal with mathematical phenomena.
- **Pragmatically:** Students learn mathematics as a response to problems in other fields.

Many researchers in Mathematics Education follow the constructivist theory of Piaget: mathematics is not something that you **have**, but it is something that you **do**. This theory fits well with much of what we know about the mathematical knowledge and mathematics learning of elementary students.

Taking the constructivist viewpoint, we must believe that students construct their mathematics and, then, we need to learn how they construct it and what they construct. Students may actually understand more mathematics than we think, but are unable to communicate their understanding within the confines of algebraic symbols. What "mathematics" have our students constructed who are entering our universities? Does the constructivist theory hold with our Calculus students? Are they constructivists or pragmatists? How have we been teaching our Calculus courses?

Answers to these questions and others will affect the way in which we teach and assess our Calculus courses. They will also affect the way in which we try to determine the effectiveness of the Calculus Reform material, as well as the effectiveness of computer mathematics systems.

The Situation at UNC Charlotte

How has *Maple* changed the way in which I teach? Dramatically! How have the Calculus Reform materials changed the way in which I teach? Also, dramatically! In fact, the two fit together quite well.

All of the Calculus Reform materials assume that the student has access to some sort of technology. Some are very specific, such as the Uhl-Porta project at the University of Illinois, closely tied to *Mathematica*. Project CALC has been closely aligned with *MathCAD* and *Derive*; the Oregon State Project, HP calculators; the Purdue Project, ISETL and *Maple*. Most of the projects are rather independent of the technology, letting the teacher choose something appropriate.

We have used both the Oregon State [DP92] material and the Harvard [HHG92] material here. We have used TI-81 calculators and *MapleV* on Sun IPC workstations networked using the afs file system. The TI-81 calculators have a very short learning curve, are programmable, and draw reasonable graphs for planar objects. Students quickly learn to use them and use them in many classes. *Maple* takes longer for them to learn and, at first, I find a number of students resistant to using and learning *Maple*. One reason that they give is that when they go on to later courses in Engineering they feel certain that they will be able to use a calculator, but they fear that they will not be able to use *Maple*. The longer they use it and the more interesting the problems to which it is applied, the more they like it. After two semesters many feel comfortable enough to continue using it in their later courses.

Using the Calculus Reform materials that are oriented toward teaching the concepts of Calculus, I spend class time teaching more concepts and fewer algorithms. *Maple* makes it possible to change the type and complexity of problems that I ask the student to solve. The problems have more realistic data and often do not have an analytical solution, or not a nice one. Relieving the students from the manipulations should enable them to spend more time planning a method of solution and interpreting the results.

Consider the following syllabus for the first year Calculus course and compare it to the standard syllabus. Note that *Maple* meshes well with the reform syllabus. This syllabus is based on using the Hughes-Hallett, Gleason, *et.al.* text. Each semester is a 14 week semester and the Calculus class meets 4 days per week, for a total of 56 class meetings.

In the first two and a half weeks we cover *A Library of Functions*. In this chapter the students get a brief review of functions and are introduced to functions as they can be represented graphically and numerically. The exponential and logarithmic functions are introduced here, as defined numerically by the computer-calculator. The students spend time becoming familiar with the graphical and numerical representations for functions. *Maple* is used here to give a wider variety of functions and graphs. Also, the students are introduced to *Maple* `procedures`, reinforcing the concept of function. In addition, the concepts of accuracy and roots are introduced. Roots are discovered approximately by zooming-in graphically, using `fsolve(...)`, or using the bisection method. We are only using the tool at this point, not explaining completely how the tool works.

The next two weeks cover an intuitive introduction to the derivative—geometric and numerical. This introduces the idea of local linearity. Using *Maple* the student will look at the symmetric difference quotient and different possible definitions of the derivative. *Maple* makes it a reasonable task to graph these different "derivative definitions" and compare them to the derivative of a function.

The next 2 weeks are then an intuitive introduction to the integral using Riemann sums and limits. *Maple* and the `student` package are helpful here. The students can use Riemann sums to compute the necessary integrals. More time is spent on the concepts associated with integration. The techniques are done later. This is called *resequencing of skills* by Heid and Judson [Heid88, Jud90].

The next 2 and a half weeks deal with the techniques of differentiation, the power rule, the product rule, the chain rule, transcendental functions and implicit differentiation. This is followed by about 2 weeks on the applications of the derivative. Note that there is no need to use the derivative for graphing as we used to do. We do spend time on graphing and

the derivative, but the focus and the questions are different.

At this point we start the integral. We do not spend a lot of time on the techniques of integration as algorithms to be learned. Instead, we study integration by parts as the reverse process to the product rule of differentiation, and integration by substitution as the reverse process to the chain rule. *Maple* and the `student` package are used here to help the students experiment. After working with it for a while, I have found the students will eventually learn to use *Maple* to experiment with different choices in these techniques. Their experimentation though has a purpose. They are trying to find rules for integration by parts and substitution that they can then memorize. They still want to classify all antidifferentiation problems so that they can then memorize the appropriate rule.

Following this we talk about the different numerical approximations of the integral: Midpoint Rule, Trapezoid Rule, and Simpson's Rule. With the technology these subjects really come alive for the students and they begin to enjoy the subject. Some prefer to approximate all integrals after that if they don't have access to *Maple*. This is something we need to work on.

Next we look at applications of the definite integral—standard fare, but done really from the standpoint of setting up the Riemann sum. Having worked with Riemann sums for most of the course, the students don't seem to dislike this as much as in traditional classes.

At this point we introduce differential equations. We use slope fields, Euler's method and Separation of Variables to handle most of the equations we encounter. The `ODE.m` package is extremely useful for *MapleV, Release 1*. The package is used by the students at their stage of development for drawing slope fields and using the `firsteuler` procedure to approximate a solution curve.[2] We ask questions about solutions that are not dependent upon finding an analytic solution. Once again the graphical and the numerical solutions are readily available to the

---

2 *MapleV Release 2* has the tools of this package incorporated into its DETools library.

student, and quite possibly more understandable than the algebraic solution.

An Example

At the time we first discuss integration, we discuss the Fundamental Theorem of Calculus. Note that none of the symbolic techniques of computing derivatives or antiderivatives have been discussed at this time. The students do know what an antiderivative is, but not having studied the symbolic representation of the derivative to any extent yet in the course. As such, they do not have a large repertoire of derivatives and antiderivatives.

The class and I work through the following in our *Maple* lab in order to deal with the concepts in the Fundamental Theorem of Calculus. We start with a function, usually something that does not have a closed form antiderivative in terms of elementary functions:

```
> f := x -> sin(x^2):
```

We then define, what we call, the "area function" based on this function:

```
> A := x -> evalf(int(f(t),t=0..x)):
```

We could use the `student` package and define the area function *via* the midpoint Riemann sum:

```
> B := x -> middlesum(f(t),t=0..x,250);
```

The students can then plot this function or get a table of values:

```
> plot(A(x), x=0..5);
> for u from 0 by .1 to 2 do
> print(u, A(u)) od;
```

The purpose is to get the students to see this "area function" as a function. Since this is a function, we then a symmetric difference quotient to compute the derivative of this function.

```
SymDiffQuot := proc(f,pt,h)
#Symmetric Difference Quotient
#Inputs
#f = function for difference quotient
#pt = point at which difference
# quotient is computed
#h = tolerance
    local sdq;
    sdq := (f(pt+h)-f(pt-h))/(2*h);
    RETURN(evalf(sdq));
end;
#
```

```
#
F := x -> SymDiffQuot(A,x,.001);
```

Once again the students make a table of values for this function and plot both this result and the original function on the same axes. The students then see that they have pretty well recovered the original function.

This is not a new idea. Here we have used the numerical and the graphical representations made possible by *Maple* to emphasize the relationship between the integral as the area under the curve and the derivative as the slope of the tangent line. Do the students understand the Fundamental Theorem of Calculus better? Anecdotally, I can say that most students do see the FTC differently. They do not perceive it as merely a means to an end. Most of the students do tend to believe that every continuous function has an antiderivative on a closed interval. Now, if they cannot find a formula for the antiderivative they will use *Maple* or a programmable, graphing calculator to plot the antiderivative over the given interval.

Conclusion

There is still a lot to be learned about the interplay between the student and the computer mathematics system. We do not know enough about how it affects the learning of mathematics. What we do know is not enough to draw sharp conclusions about the appropriate usage of a CMS in a Calculus course. We do know something from the research [Hart92]:

- Students using the calculus reform material showed greater facility with graphical and numerical representations.
- Individual students have a definite preference for one particular representation, but different factors determine the preference.
- Students' use of computer mathematics systems is closely tied to their management of the representations.
- When a device (CMS or even a formula) is used to perform a computation in a **routine** manner, the students look at this result less critically.
- Students' confidence in the graphical information from the computer mathematics system is closely tied to having *a priori* information.
- Student attitudes towards calculus improved when using the Calculus Reform material.

- Student attitudes toward mathematics improved when using a CMS, regardless of the particular calculus curriculum.

There is much that a computer mathematics system can offer a calculus student. However, if this is not done in conjunction with a curriculum which takes advantage of the different strengths of the computer, we will not be having as much of an effect on the student. The computer mathematics system can offer a lot to the calculus class, but we must never look on it as a panacea for all of our ills.

References

[Beck90]   Charlene E. Beckmann. Effect of computer graphic use on student understanding of calculus concepts. In Franklin Demana, Bert K. Waits, and John Harvey, editors, *Proceeding of the Conference on Technology in Collegiate Mathematics*, pages 104–107, Reading, MA, 1990. Addison Wesley

[Dev90]   J.S. Devitt, Adapting the Maple computer algebra system to the mathematics curriculum. In Franklin Demana, Bert K. Waits, and John Harvey, editors, *Proceeding of the Conference on Technology in Collegiate Mathematics*, pages 12–27, Reading, MA, 1990. Addison Wesley

[Doug86]   Ronald G. Douglas, editor. *Toward a Lean and Lively Calculus*, Washington DC, 1986. The Mathematical Association of America. MAA Notes 6.

[DP92]   Thomas P. Dick and Charles Patton. *Calculus*, Volume I. Wadsworth Publishing Co., preliminary edition, 1992.

[Dub92]   Ed Dubinsky. A learning theory approach to calculus. In Zaven A. Karian, editor, Symbolic Computation in Undergraduate Mathematics Education, pages 43–55, Washington, DC, 1992. The Mathematical Association of America.

[Hart92]   Dianne Hart. Building concept images—supercalculators and students' use of multiple representations in calculus. Ph.D. thesis, Oregon State University, 1992. Thomas P. Dick, advisor.

[Harv90]   John Harvey. changes in pedagogy and testing when using technologies in college-level mathematics courses. In Franklin Demana, Bert K. Waits, and John Harvey, editors, *Proceeding of the Conference on Technology in Collegiate Mathematics*, pages 40–54, Reading, MA, 1990. Addison Wesley

[Heid88]   M. Kathleen Heid. Resequencing skills and concepts in applied calculus using the computer as a tool. *Journal for Research in Mathematics Education*, 19(1):3–25, 1988.

[HHG92]   Deborah Hughes-Hallett, Andrew Gleason, *et. al. Calculus*, John Wiley, New York, NY, preliminary edition, 1992.

[HLLL92]   Joel Hillel, Lesley Lee, Colette LaBorde, and Liora Linchevski. Basic functions through the lens of computer algebra systems. *Journal of Mathematical Behavior*, 11:119–158, 1992.

[HSM90]   M. Kathleen Heid, Charlene Sheets, and Mary Ann Matras. Computer-enhanced algebra: New roles and challenges for teachers and students. In Thomas J. Cooney and Christian R. Hirsch, editors, *Teaching and Learning Mathematics in the 1990s: 1990 Yearbook*, Chapter 23, pages 194–204. National Council of Teachers of Mathematics, Reston, VA, 1990.

[Jud90]   Phoebe T. Judson. Elementary business calculus with computer algebra. *Journal of Mathematical Behavior*, 9(2):153-157,1990.

[Kar92]   Zaven Z. Karian, editor. *Symbolic Computation in Undergraduate Mathematics Education*, Washington DC, 1992. The Mathematical Association of America. MAA Notes 24.

[Palm91]   Jeanette Palmiter. Effects of computer algebra systems on concept and skill acquisition in calculus. Journal for Research in Mathematics Education, 22(2):151–156, 1991.

[Ral90]   Anthony Ralston. the effect of technology on teaching collegiate mathematics. In Franklin Demana, Bert K. Waits, and John Harvey, editors, *Proceeding of the Conference on Technology in Collegiate Mathematics*, pages 78–82, Reading, MA, 1990. Addison Wesley

[SH86]   Donald B. Small and John M. Hosack. Computer algebra systems, tools for reforming calculus instruction. In Ronald G. Douglas, editor, Toward a Lean and Lively Calculus, pages 143–156, Washington DC, 1986. the Mathematical Association of America. MAA Notes 6.

[Zorn90]   Paul Zorn. Algebraic, graphical and numerical computing in elementary calculus: Report of a project at St. Olaf College. In Franklin Demana, Bert K. Waits, and John Harvey, editors, *Proceeding of the Conference on Technology in Collegiate Mathematics*, pages 92–95, Reading, MA, 1990. Addison Wesley

David C. Royster received his B.A. in mathematics from the University of the South, Sewanee, TN in 1973 and his Ph.D. from Lousiana State University in 1978. His thesis was written under the direction of Pierre Conner. His research interests are group actions on manifolds, differential topology and geometry. After teaching at the University of Virginia and the University of Texas, Austin, he is currently an associate professor of mathematics at the University of North Carolina at Charlotte. Current research involves research into the learning of calculus and the use of computer mathematics systems in college level courses. He can be reached at:

Department of Mathematics
UNC Charlotte
Charlotte, NC 28223
USA
email: droyster@unccsun.uncc.edu

INTERACTIVE MATHEMATICS TEXTS: IDEAS FOR DEVELOPERS

Carol Scheftic
Department of Mathematics, Carnegie-Mellon University, Pittsburgh PA, USA

Abstract

If you're going to try and reinvent the gear, it may be easier if you don't have to reinvent the wheel first. If you're going to try and develop interactive mathematics texts (IMTs) using Maple, it may be easier if you have experience as a mathematician, teacher of mathematics, computer programmer, instructional designer, developer of computer-based instructional materials, and textbook author, in addition to being familiar with a computer algebra system such as Maple.

Few IMT developers have experience in all those areas, and there is no comprehensive theory to guide those entering the field of IMT design and development. For those who like interesting challenges, this presents the opportunity to develop additional skills and to begin to formulate just such a theory. In this paper I will describe a collection of "wheels," important basic ideas from each of the relevant disciplines, demonstrate how they can be incorporated into a Maple-based IMT, and illustrate what can go wrong if these suggestions are ignored. The "wheels" to be included will be chosen from those that I have found to be general enough to be useful across a number of different IMT applications and settings.

Mathematical Computation with Maple V:
Ideas and Applications
Tom Lee, Editor
©1993 Birkhäuser Boston

Introduction

Interactive texts are a new medium, one in which IMT authors, math teachers, and students alike have little experience. It will take time, experimentation, and practice before IMTs will come into their own. Early films started out being little more than recorded stage plays, yet the film industry today is significantly larger and more influential than is the theater. It makes sense that our first attempts at IMTs will mimic textbooks, even though textbooks evolved from lectures and both of those are targeted toward groups. It is nonetheless critical to realize that much of the strength of an IMT derives from the interactivity of its medium and its opportunity for extensive individualization.

If you want to develop your own interactive texts, you should be familiar with a variety of interactive products, both within your own area and across many other disciplines. Take time to look at drill and practice software, at tutorial software, at interactive multimedia, even at computer games. Look for features that make them easy to use, that lead to confusion or frustration, that seem to use the medium appropriately.
Look for differences in style that depend on the goal of the activity, such as theory, practice, application, or interpretation.

The following sections will cover a number of such characteristics that arise often, particularly for beginning IMT developers; as you gain experience, you will be able to extend this list yourself.

Visual Displays

Chabay and Sherwood (1992) emphasize that the display is central in on-line materials, and thus tied into almost all design decisions. While we can use it to display text, graphics, and mathematical information, we should remember how it differs from our more traditional methods of presentation. Blank space, for example, is cheap on-line, so it can and should be used liberally.

When writing on-line materials, you should direct the reader by using a variety of layout details in place of the denser format you would use for a printed document. Even a highly motivated person with good concentration will find that easier to follow than full paragraphs of text on a display screen. Compare this paragraph with the following:

```
When presenting text on-line:
   -  avoid paragraphs,
   -  guide the reader's attention,
   -  set off key ideas.
Compare this format with the paragraph
shown above.
```

Asking versus Telling

It takes time to develop a writing style that is clear and smooth yet also encourages interaction. We are used to writing in an expository style, and an interactive style is more demanding of the author. It is more demanding of the students as well, which is actually an improvement over the traditional static presentation.

One guideline for authors new to interactive text is to try and think about your expectations for the students with each phrase you write. Whenever there is a place where you want the reader to stop and think about something, to make a connection to something that has been done before, to make an inference about what will come next, there your prose should stop and you should insert a question. In some cases, you will then want your software to evaluate the student's reply; in others, you can follow the question-and-answer sequence with additional information that the student can use for self-evaluation. Try to be polite and friendly throughout your writing, but avoid being overly "cute" in it.

Use of Graphics

When writing interactive text materials, make liberal use of the computer's graphic display capabilities. Mix text and graphics throughout: use pictures to illustrate information from the text, and use text to clarify information from the pictures. But don't try to crowd an overwhelming amount of information into a visual display. As with text, you are free to use white space liberally throughout your graphics. Use a series of clearly labeled images to illustrate a collection of ideas; repeat them in an overlaid fashion only at the end, if appropriate, in order to present a final comparison or contrast among them.

Use of Color

Be careful in your use of color. First of all, make sure that you avoid implying spurious correlations. If you use a particular set of colors, for example, then what does any particular one mean? While you may just like a particular shade of blue and think that its appearance doesn't mean anything beyond that, your students may assume, incorrectly, that blue will always represent the derivative or the error term or something else that will eventually lead them astray. Secondly, be extremely wary of using colors to represent specific objects or processes. You can't be sure that the students will be able to see the same colors that you see: they may be working on a machine with a monochrome monitor, or one with fewer or different color choices, or they may even be working on a color machine but be color-blind themselves. While the use of color coding to carry information can be extremely powerful, you should also be sure to give a color key early on for reference and to describe the relevant distinction in text. Until you gain experience, test your work occasionally using a monochrome display to see whether you have documented all the color-critical information you have included.

Use of Labels

Whether or not you use different colors to convey information in a plot, it is always a good practice to label each curve with a text description. All axes should be clearly labeled as well. This is a practice you should model for your students in the materials you write, and it is a reasonable for you to require them to do the same in the materials they submit. Illustrations or other graphic images imported from a drawing program should also be clearly labeled.

Inconsistencies among Display Media

Some software displays are comparable when shown on different monitors and when printed. Some look similar, but not identical. Others appear extremely different. You need to know what your software does, and modify your design accordingly. If you highlight certain areas of text or reposition text or objects in order to refer to specific elements, make sure that these features are carry through on the various display media that your students will use.

Interface versus Content

Teach the interface independently of the content of a lesson. Most first-time developers, as they learn to use a system and develop their first lesson materials in it, have a tendency to mingle interface with content. This is, in fact, one way to get yourself familiar with the software and authoring in it all at once, but don't expect such a lesson to be successful when used by your students. A much better strategy is to teach the interface while covering material that students already know. Use that as an opportunity for review! Once they are comfortable with the interface, then you can have them work with it for covering new material.

Interaction

Keep the materials as interactive as possible. Ask, don't tell, the students what is going on. Keep them working in an active mode. Avoid letting them become passive. It's fine to break up some passages to fit the screen by having them press the return key, but use that approach sparingly. Ask questions that require them to think about the material as they work their way through it.

Provide thorough instructions, but keep them as terse and telegraphic as possible. Don't merely tell the students what to do: first, show it, and then ask them to perform a similar task themselves. Keep instructions near the action. On the other hand, don't let your lesson introduction get so bogged down in instructions that the students become totally bored before they ever get to the major points. Start off simply but directly, and add components as necessary to build up the required sequences.

Input and Output

While you want to keep the students active, you also want to minimize the typing that you expect them to do. While you do want to require more of the students than simply reexecuting your input sequences, it is reasonable to model a set of commands that require slight modification by the student. Then, depending on the situation, students may simply modify an initial statement and repeat the sequence of commands that follow, or they may copy the entire sequence into their own file and make a larger set of modifications in answering some question that you have posed. As they become more familiar with both the package and the subject matter, you can gradually increase the amount of actual response construction, with all the attendant typing, that they must do.

Input/Output Plus Text

The use of modern CASs involves more than just input and output. Their interfaces permit the use of textual information. In addition to using that for your own presentation, require students to use it as part of their answers. Although it is challenging for students to write, it is now quite easy for them to include. Although it is challenging for instructors to grade, it is a powerful method for determining students' understanding. Since this will be a new requirement, particularly in a mathematics course, for many students, it is very important that you model the expected behaviour. As much as possible, extend the learning that takes place by allowing, even encouraging, revision. Help students to learn that the issue is not whether the answer is right or wrong the first time. Instead, they need to learn that a far more important skill is to critically review what they have done, and continue to revise it until they do get it right and understand why that answer is acceptable.

Additional Hints and Strategies

A series of additional hints for developers can be summarized as follows:
- Explore the medium.
- Try new instructional approaches.
- Develop interactively.
- Move from the specific to the general.
- Start at the mid-range of difficulty, and adjust as necessary.
- Test and observe.
    Revise.
    Retest.
    Repeat.
- Be flexible.
- Have fun.

Don't take on too large a project from the start. Instead, pick at most two or three topics that seem most suited to this new approach and experiment with those. Working with more than one topic can provide you with a way to compare and contrast the use of different approaches for different topics. Limiting yourself to a small handful permits you to try different things without worrying so much about sticking with a consistent style. Discover approaches that you find comfortable with,

and then both observe and talk with your students about their reactions. Don't spend so much time on any one lesson in the beginning that you will be hesitant to throw it away and start over for the next time. Every top software developer can tell you about wonderful ideas that they had to scrap when they discovered their limitations in practice. With time, you'll find your voice in this new medium, you'll find collaborators with compatible styles, and you'll be ready to take on an even larger project.

One Final Reminder

As you work, remember to keep asking yourself this question: If the machines can compute the answers, what is it that the students need to learn?

SELECTED BIBLIOGRAPHY

Alessi, Stephen M. & Stanley R. Trollip. (1985). *Computer-Based Instruction: Methods and Development.* Englewood Cliffs, NJ: Prentice-Hall.

Barwise, Jon, & John Etchemendy. (1989). Creating Courseware. In *Notices of the AMS,* Vol. 36, No 1, pp. 32-40.

Bork, Alfred. (1985). *Personal Computers for Education.* New York, NY: Harper & Row.

Bork, Alfred. (1987). *Learning with Personal Computers.* New York, NY: Harper & Row.

Brown, Stephen I., & Marion I. Walter. (1983). *The Art of Problem Posing.* Hillsdale, NJ: Lawrence Erlbaum Associates.

Chabay, Ruth, & Bruce A. Sherwood. (1992). A Practical Guide for the Creation of Educational Software. In Larkin, Jill H., & Ruth W. Chabay (Eds.), *Computer-Assisted Instruction and Intelligent Tutoring Systems: Shared Goals and Complementary Approaches.* Hillsdale, NJ: Lawrence Erlbaum Associates.

Ericsson, K. Anders, & Herbert A. Simon. (1984). *Protocol Analysis: Verbal Reports as Data.* Cambridge, MA: The MIT Press.

Gallini, Joan K., & Margaret E. Gredler. (1989). *Instructional Design for Computers: Cognitive Applications in BASIC and Logo.* Glenview, IL: Scott, Foresman.

Gardner, Howard. (1983). *Frames of Mind: The Theory of Multiple Intelligences.* Basic Books.

Heines, Jesse M. (1984). *Screen Design Strategies for Computer-Assisted Instruction.* Bedford, MA: Digital Press.

Johnson, D. W., & F. P. Johnson. (1987). *Joining Together: Group Theory and Group Skills (Third Ed.).* Englewood Cliffs, NJ: Prentice-Hall.

Johnson, D. W., & R. T. Johnson. (1975). *Learning Together and Alone: Cooperation, Competition, and Individualization.* Englewood Cliffs, NJ: Prentice-Hall.

Kearsley, Greg. (1986). *Authoring: A Guide to the Design of Instructional Software.* Reading, MA: Addison-Wesley.

Keller, Arnold. (1987). *When Machines Teach: Designing Computer Courseware.* New York, NY: Harper & Row.

Larkin, Jill H., & Ruth W. Chabay (Eds.). (1992). *Computer-Assisted Instruction and Intelligent Tutoring Systems: Shared Goals and Complementary Approaches.* Hillsdale, NJ: Lawrence Erlbaum Associates.

Moursund, Dave. (1986). *Computers and Problem Solving: A Workshop for Educators.* Eugene, OR: International Society for Technology in Education.

Robbat, Richard J. (1985). *Computers and Individualized Learning: Moving to Alternative Learning Environments.* Eugene, OR: International Society for Technology in Education.

Schoenfeld, Alan H. (1987). *Cognitive Science and Mathematics Education.* Hillsdale, NJ: Lawrence Erlbaum Associates.

Shneiderman, Ben. (1987). *Designing the User Interface: Strategies for Effective Human-Computer Interaction.* Reading, MA: Addison-Wesley.

Steinberg, Esther R. (1991). *Computer-Assisted Instruction: A Synthesis of Theory, Practice, and Technology.* Hillsdale, NJ: Lawrence Erlbaum Associates.

Walker, Decker F., & Robert D. Hess. (1984). *Instructional Software: Principles for Design and Use.* Belmont, CA: Wadsworth.

Zimmerman, Walter, & Steve Cunningham (Eds.). (1991). *Visualization in Teaching and Learning Mathematics.* MAA Notes Number 19.

About the Author

Carol Scheftic has a B.S. in Mathematics from Carnegie-Mellon University (1971), and both an M.A. in Mathematics Education (1973) and a Ph.D. in Educational Technology (1985) from the University of Pittsburgh. In the process of researching appropriate applications of technology in education, she has taught mathematics and computer science at all levels from elementary school through post-graduate professional education. She is currently a Lecturer with the Department of Mathematics at Carnegie Mellon, where she directs projects on collaborative learning and the uses of computer algebra systems in undergraduate mathematics. She also teaches workshops for collegiate mathematics faculty via the MAA's Interactive Mathematics Text Project. She will spend the 1993-94 academic year as Associate Technology Editor with Brooks/Cole Publishing, where her responsibilities will include interactive texts.

Appendix A: CAI vs. CAS

A number of additional guidelines, appropriate for CAI, are much more difficult to implement in traditional CAS worksheets that support neither individual "page" formats nor exact placement on the screen. Depending on the sophistication of your students and the complexity of your topic, you may want to consider writing your instructional materials in a programming language that does support such features, and using it to make calls to a CAS as needed. Guidelines for such programs will not be treated in detail in this paper, but a few important suggestions from the paper by Chabay and Sherwood (1992) will be included to give some insight into the further issues involved in this type of project.

Display issues include:
- Build displays up piece by piece.
- Maintain visual context, but avoid clutter.
- Clearly make connections between text and graphics(e.g., use arrows)
- Balance adjacency versus consistency.
- Balance consistency versus variety.

Input/Output issues include:
- Avoid mixing input modes (e.g., use keyboard *or* mouse).
- Assist the student's concentration by avoiding distractions.

Interaction issues include:
- Consider validity. Anticipate common syntactic and semantic errors.
- Allow synonyms. Mathematical or verbal.
- Ignore blank inputs. But do check for a student who hasn't a clue what to do.
- Avoid browsing. Require more than just pressing the return key.
- Give clear, explicit feedback.
- Avoid dead ends.

Appendix B: Sample Interactive Text for an Introductory Calculus Course.
This is the first lesson in the course. It is used to review algebra, not to introduce calculus topics themselves. This sample is fully executed, i.e., after a student has worked through it.

Computer Algebra Systems (CASs) typically have three types of features:
1. numeric,
2. symbolic, and
3. graphic.

In this lesson, you will get a brief introduction to each of them. Throughout this course, you will learn more about the specific features that are relevant to your first-year calculus course. In addition, you should learn enough about Maple itself, how to navigate through its help files, and how to move back and forth between "textbook mathematics" and "computer algebra systems" that you will be able to pick up additional skills, as needed, for later courses in mathematics, science, engineering, computer science, etc.

Numeric Capabilities

To begin with, let's just use Maple as a simple calculator. If necessary, use the mouse to put the insertion point on the following input command (anywhere after the round "bullet" or "input prompt" but it does not have to be at the end of the line):

• 3 + 5;

$$8$$

Notice these aspects of the above sequence:
- a.   the input line finished with a semi-colon,
- b.   the output was centered on the line below the input.
- c.   if you place the insertion point on either the input or the output line, then the "mode" buttons in the will ruler change to reflect that;
- b.   if you place the insertion point inside this outline, then it will indicate that this is text.

Now, try executing the following couple of commands:

• 1/5 + 3/7;

$$\frac{22}{35}$$

• 2^100;

$$1267650600228229401496703205376$$

Notice that those answers are EXACT. Maple will always give a result in the most accurate form possible, unless you ask for a decimal approximation (using "scientific notation" if necessary). In Maple-terms, ask it to evaluate the answer in floating point arithmetic:

• evalf(1/5 + 3/7);

$$.6285714286$$

• evalf(2^100);

$$.1267650600*10^{31}$$

In the above examples, the answers are given with ten significant digits. That 10 is the default value for the number of "Digits" that Maple will use. If necessary, you can specify that Maple should use more or less than ten:

• evalf(1/5 + 3/7, 5);

$$.62857$$

• evalf(2^100, 20);

$$.12676506002282294015*10^{31}$$

If you want to experiment, enter one or more commands at the following input prompt. If you want more prompts, put the insertion point in an appropriate place and choose **Insert Prompt** from the **Format** menu.

•

ASK if you get any results you don't expect. When you're done experimenting, continue below.

Maple also uses the traditional order of operations:
1. exponentiation (either ^ or **) first,
2. then multiplication (*) and division (/),
3. then addition (+) and subtraction (-)

unless you use parentheses to indicate a particular order. It doesn't hurt to put in extra parentheses (except for the pain of unnecessary typing).

Try each of the following, and make sure you understand Maple's response to each one. ASK if any of them seem surprising to you.

- 1/2^5;

$$1/32$$

- 3/2^5;

$$3/32$$

- (3/2)^5;

$$\frac{243}{32}$$

- 1.5^5;

$$7.59375$$

What happened in that last example? See if you can figure it out before reading further.

Since the input used a floating point number (1.5, not 3/2), Maple completed the computations in floating point. It didn't use the "default" value of ten digits in the output, however, since only five are appropriate. Consider these:

- (1.5)^2;

$$2.25$$

- (1.5)^6;

$$11.390625$$

- (1.5)^8;

$$25.62890625$$

- (1.5)^12;

$$129.7463379$$

This example also shows us that there are two ways to force floating point evaluation: through use of the evalf() command and use of floating point numbers.

- 1.5^15;

$$437.8938905$$

- ((3.0)/(2.0))^15;

$$437.8938905$$

- (3./2)^15;

$$437.8938905$$

- evalf((3/2)^15);

$$437.8938904$$

Of course, when you are using parentheses, you should make sure that they match up correctly (two of the next three input commands SHOULD generate an error message):
* evalf(3/2 ^15);

$$.00009155273438$$

* evalf(3/2)^15);
syntax error:
evalf(3/2)^15);
           ^
* evalf((3/2 ^15);
syntax error:
evalf((3/2 ^15);

          ^

In each of the above cases, a space was left on the input line (as a clue for you) where the missing parenthesis(es) should be.

Notice the difference in the error messages:
    *   in the first case, the caret points to a right parenthesis
       indicating that there was no opening parenthesis that will match it, but
    *   in the second case, the caret comes at the end,
       indicating an open right parenthesis that wasn't closed before the final semicolon.

While Maple doesn't know what you intended to do, so it may not actually point to the place where your error really is, its error message should still give you some useful information. With experience, you'll learn how to interpret them yourself. For now, you can either retype the command yourself (or copy, paste, and edit it) at the prompt provided below. Alternatively, you can just go up to either of the previous lines, insert an appropriate parenthesis, and re-execute that command.
*

Maple's Help System

One of the real advantages of using Maple is that it has a very useful and extensive Help system. We will demonstrate it for you in class, and we are providing a little summary of its features in the Hints document. But if you ever want to learn more about a command that is introduced in one of these lessons, you can read the relevant help file. For example, execute the next input line to read more about the "evaluate using floating point arithmetic" command:
* ?evalf
*

Symbolic Capabilities

While those numeric capabilities are useful, they are certainly not the only reasons for using a CAS. Maple can also operate with symbols, as well as numbers:
* factor(x^2 - 5*x + 6);

$$(x - 2) \ (x - 3)$$

* solve(x^2 - 5*x + 6 = 0, {x});

$$\{x = 3\}, \ \{x = 2\}$$

Some of Maple's symbolic commands also have default values that can save you some typing. In the case of solve(), if there's only one variable, Maple will solve for that one. If you just give an expression, it will solve for when that is equal to zero. Thus, the last command could have been written as follows (but do notice the difference in the format of the output).

• solve(x^2 - 5*x + 6);

$$3, \ 2$$

One of the most powerful features of Maple, however, is its ability to do fully symbolic manipulations. For example, see if you recognize the result when you give Maple a "generic" quadratic equation to solve:

• solve(a*x^2 + b*x + c = 0, x);

$$1/2 \ \frac{-b + (b^2 - 4\,a\,c)^{1/2}}{a}, \ 1/2 \ \frac{-b - (b^2 - 4\,a\,c)^{1/2}}{a}$$

Maple can also solve systems of equations. We'll do one simple set next. Later this term, we'll learn two more methods that will be useful with larger systems (one uses this method, with some typing shortcuts, and the other uses techniques from linear algebra).

To set up a system of m equations in n unknowns, is it useful to assign each equation to a name.

A name in Maple:
* must begin with a letter,
* can consist of anywhere from 1 to 499 characters, and
* after the first character, it may contain either letters or numbers.

Notice that Maple is sensitive to case:
* a lower-case x and an upper-case X are treated as two distinct characters.

• eq1 := -2*x +   y - 3*z = 1;

$$eq1 := -2\,x + y - 3\,z = 1$$

• eq2 := 2*x - 2*y +   z = -3;

$$eq2 := 2\,x - 2\,y + z = -3$$

• eq3 :=   x +   y -   z = -3;

$$eq3 := x + y - z = -3$$

Now that the equations have been set up, we can use the solve command. Notice that both the list of equations and the list of variables to be solved for are enclosed in pairs of braces:

• solve({eq1, eq2, eq3}, {x, y, z});

$$\{y = 0, \ z = 1, \ x = -2\}$$

The answers may not have come out in the order you expected (i.e., they may not be listed as x, y, z, in that order), but they should all be there. Notice the difference between the above and the situation where you use fewer equations than you have unknowns:

• solve({eq1, eq2}, {x, y, z});

$$\{y = y, \ z = -1/2\,y + 1, \ x = 5/4\,y - 2\}$$

A solution of the form y=y means that the value of y is arbitrary (it can be assigned any value). Then the other variables, such as x and z, are specified in terms of the chosen y.

•

Graphical Capabilities

Students quickly learn, in first-semester Calculus, that curve sketching can be a very time-consuming activity. On the other hand, an image of the function or expression one is working with can often lead to valuable insights. Luckily, computer algebra systems such as Maple are extremely helpful in this area.

The simplest form of the plot command just uses Maple's defaults for everything, as illustrated with the following input. After you have executed the following command, arrange your windows so that you can see this input, the text that follows it, and your plot, all at the same time. (ASK if you need help in getting everything visible at once.)
• `plot(x * sin(x));`

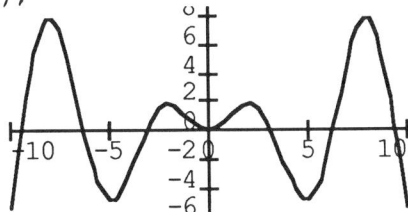

(Students executing this file will see a large plot in its own windoe, rather than the above.)

Notice that the plot defaults include:
* horizontal axis from -10 to 10,
* vertical axis scaled to whatever Maple thinks is appropriate for the graph,
* tickmarks on the axes,
* no labels anywhere.

(When done looking at the plot, click in this window to make it active again, and continue with the lesson.)

•

Since we are going to plot this same expression, x*sin(x), a number of times, let's give it a name and then save ourselves some typing by just using that name from now on.
• `f := x * sin(x);`

$$f := x \sin(x)$$

For our first improvement on the plot, let's have the horizontal axis go from -3Pi to 3Pi.
• `plot(f, -3*Pi..3*Pi);`

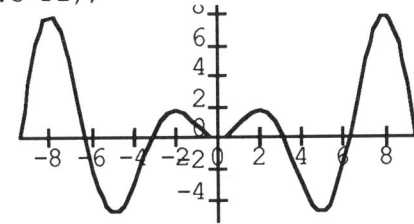

Next, let's make it clear that "x" is the variable represented by the horizontal axis:

- `plot(f, x=-3*Pi..3*Pi);`

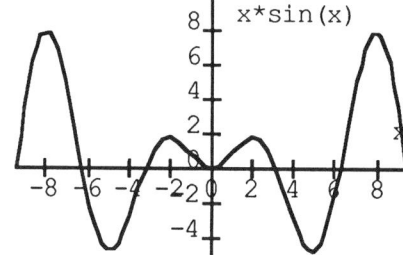

Note: For this course, you must provide AT LEAST the above information for every plot you submit for grading. That is, in addition to giving the function you want Maple to plot, you must indicate at least the variable being plotted and select an appropriate interval yourself.
-

Hint: You will start getting "Out of Memory" warnings or errors if you leave a lot of plot windows open. It's fine to have a few open at once, particularly if you want to compare the differences between the results of two different plot commands.

When you are done with a particular plot, close its window.

You can always reexecute the relevant plot command to see it again later!

To close a window when it is active (i.e., when it is "in front" and there are horizontal lines in its title bar), you can:
   a. go to the **File** menu and choose **Close**, or
   b. hold down the **Cloverleaf** (aka **Command**) key and press the letter **w**, or
   c. click in the Close-Box at the far left of the plot's title bar.
-

Back to our function. Let's indicate that the function <u>x*sin(x)</u> is shown on the vertical axis.

Note: The *back-quotes*, from the key next to the numeral 1 in the upper left corner of the keyboard, are *required* because the string contains non-alphabetic characters such as the asterisk and parentheses.
- `plot(f, x=-3*Pi..3*Pi, `x*sin(x)`);`

If you want, you can specify a range for the vertical axis too:
- `plot(f, x=-3*Pi..3*Pi, `x*sin(x)`=-5..10);`

Finally, we can give this plot a title:
```
• plot(f, x=-3*Pi..3*Pi, `x*sin(x)`=-5..10, title=`My first labelled
plot.`);
```

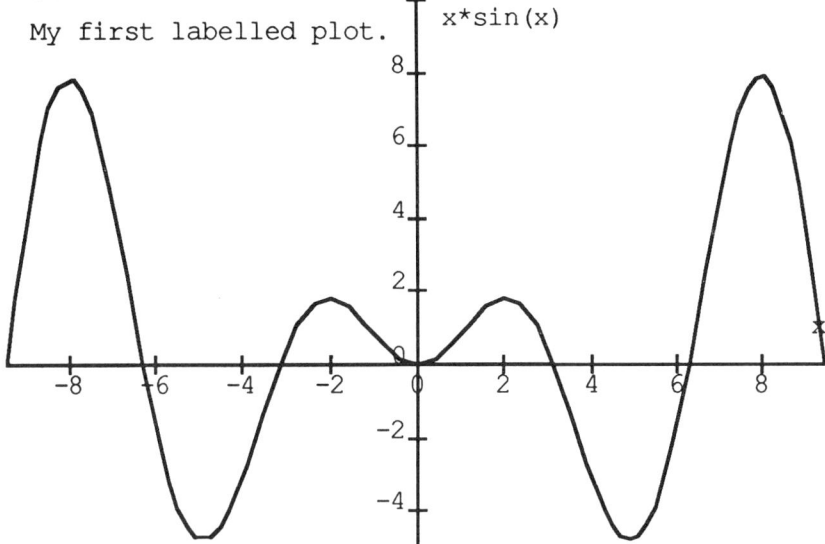

(This plot was copied from the default-size plot window.)

Notice that the stated information "just fits" in the default-size plot window. The amount of information you can conveniently display on a single plot is partly a factor of the anticipated display format.

- If you enlarge the plot window (by clicking in its Zoom-Box, at the far right end of the plot window's title bar), then everything still fits, and the title will still be centered along the top of the second quadrant.
- If you reduce the plot window (by dragging up on the resize box in the bottom right of the plot window), however, and make it smaller than the default, then the title no longer fits in the upper left quadrant. Maple will then center it, overwriting the label on the vertical axis.

Note: For more information on Maple's plotting routines, look up the help files on plot, plot3d, plot[options], and other files mentioned within those sources of information.
•

When you have worked your way through all of this, Quit Maple, and then restart it via the file Problems #01. If you want, you can reopen this file again (after you have the assignment opened as the primary Worksheet) and use this one as a Scratchpad.

II

MAPLE V IN MATHEMATICS

TRUNCATION AND VARIANCE IN SCALE MIXTURES

William C. Bauldry, Jaimie L. Hebert
Dept. of Mathematical Sciences, Appalachian State University, Boone NC, USA

1. Introduction

Let X be a continuous nonnegative random variable with density f, mean μ, and finite variance σ^2. Mullooly (1988), hereafter simply Mullooly, has shown that if f is positive on the interior of its support, $\lim_{x \to 0} f(x) > 0$, and $\frac{\sigma}{\mu} > 1$, then σ^2 may be increased by truncation. Denote by $\sigma^2(t)$, the variance of the truncated random variable $X_t \equiv X \, I_{(t, \infty)}(X)$, where I_A is an indicator on the set A. Specifically, Mullooly demonstrates that for densities satisfying these conditions, there exists a real number $T > 0$ such that $\sigma^2(t) > \sigma^2$ for all $t \in (0, T)$. We shall call T the *variance inflation boundary* for X_t. When $\sigma^2(t) > \sigma^2$ for all $t \in (0, \infty)$, we say that $T = \infty$.

Mullooly notes that if T is finite, then it is given by the smallest positive solution of

$$\sigma^2(T) = \frac{\mathfrak{I}_2(T)}{\overline{F}(T)} - \left[\frac{\mathfrak{I}_1(T)}{\overline{F}(T)}\right]^2 = \sigma^2, \quad (1.1)$$

where $\mathfrak{I}_j = \int_t^\infty x^j f(x) \, dx$. Although (1.1) can be solved approximately for any density satisfying the conditions of his theorem, he suggests an algorithm which provides a lower bound on the variance inflation boundary, T, which is calculationally simpler. He shows that the variance of the truncated random variable, X_t, and its derivative can be represented as

$$\sigma^2(t) = 2W(t) + M^2(t)$$

and

$$\frac{d\sigma^2(t)}{dt} = 2h(t) \, W(t),$$

where

$$W(t) = \left(1/\overline{F}(t)\right) \int_t^\infty \left[x - \left(\mathfrak{I}_1(t)\overline{F}(t)\right)\right] \overline{F}(x) \, dx,$$

$$M(t) = \left(1/\overline{F}(t)\right) \int_t^\infty \overline{F}(x) \, dx,$$

and $h(t) = f(t)/\overline{F}(t)$, the hazard function. It is easily verified that $\sigma^2(0) = \sigma^2$ and $\lim_{t \to 0} \left[d\sigma^2(t)/dt\right] > 0$. Thus, the algorithm is based on the fact that $\sigma^2(t)$ must achieve a maximum before taking the value σ^2. Using these expressions, one may show that a lower bound for T is the $(1 - \frac{1}{CV})^{th}$ percentile of the distribution, where $CV = \frac{\sigma}{\mu}$, the coefficient of variation.

Barlow and Proschan (1975) note that lifetime distributions with continuous densities and decreasing hazard rates satisfy these conditions. Such distributions are quite useful in reliability studies. Of particular interest are lifetime distributions possessing *completely monotone* hazard rates. (A function h is said to be completely monotone if $(-1)^n \frac{d^n}{dx^n}\left[h(x)\right] \geq 0$ for $n = 1, 2, \ldots$, assuming these derivatives exist.) Jewell (1982) has shown that complete

Mathematical Computation with Maple V:
Ideas and Applications
Tom Lee, Editor
©1993 Birkhäuser Boston

monotonicity is a sufficient condition for a lifetime distribution to be represented as a scale mixture of exponentials. That is, if f is completely monotone, then there exists a distribution function G such that

$$f(x) = \int_0^\infty \lambda \exp(-\lambda x) dG(\lambda). \qquad (1.2)$$

We denote by $\mathcal{F}$ the set of densities which can be written in the form (1.2). From the discussion above, it is apparent that every $f \in \mathcal{F}$ will satisfy the conditions of Mullooly's result.

Hebert and Seaman (1994), hereafter simply Hebert and Seaman, examine the difference, $\sigma^2(t) - \sigma^2$, relative to the mixing distribution $G(\lambda)$. They provide a general form for the difference in variances, $\sigma^2(t) - \sigma^2$, in terms of the mixing distribution. Specifically, let X have density $f \in \mathcal{F}$ and corresponding mixing distribution $G(\lambda)$, let $X_t = X I_{(t,\infty)}(X)$ and let $Y = X_t - t$. They show that the distribution of Y can be represented as a mixture of exponential densities with mixing distribution

$$G^*(\lambda) = \int_0^\lambda H_u(t) \, dG(u) \qquad (1.3)$$

where $H_u(t) = \dfrac{\exp(-ut)}{\mathcal{L}_t\{G(u)\}}$ and $\mathcal{L}_t\{G(u)\}$ is the Laplace-Stieltjes transform of G with frequency variable t. Since Y simply represents a change in location in X_t, the variance of Y is also σ^2. Thus, the difference in variance may be represented as

$$\begin{aligned} \sigma^2(t) - \sigma^2 &= 2 \, \mathrm{E}_\Lambda\left[\Lambda^{-2}\big(H_\Lambda(t) - 1\big)\right] \\ &\quad - \mathrm{E}_\Lambda\left[\Lambda^{-1}\big(H_\Lambda(t) + 1\big)\right] \\ &\quad \cdot \mathrm{E}_\Lambda\left[\Lambda^{-1}\big(H_\Lambda(t) - 1\big)\right], \qquad (1.4) \end{aligned}$$

where $\mathrm{E}_\Lambda[\cdot]$ denotes expectation with respect to

$G(\lambda)$. Clearly, T is the smallest positive solution of (1.4) when in fact T is finite.

The difference in variances may be derived directly in many instances. The value of (1.4) is that the difference is obtained by averaging over the hazard rates, i.e., each of the expected values in (1.4) is with respect to G. Such a representation is useful when studying the sensitivity of this difference relative to changes in the mixing distribution. Since an exponential mixture is uniquely identified by the corresponding mixing distribution, characteristics of the variance of that mixture are dependent on parameter specifications in the mixing distribution. In this sense, information concerning the behavior of the stochastic hazard rate will provide information concerning the variance of the mixture. Information concerning the change in variance, relative to the behavior of G, is provided through (1.4).

Hebert and Seaman consider several examples in their manuscript. The variance inflation boundary, when finite, is provided for each of these examples. In several of these examples, the authors compare the actual value of T to the lower bound provided by Mullooly's algorithm. In the present manuscript, we illustrate the use of Maple in deriving the calculations in the examples of Hebert and Seaman and extend the results of their discrete mixture example. By plotting expression (1.4) as a function of the mixing parameters, we investigate the sensitivity of the variance of the truncated random variable relative to these parameters.

2. A discrete mixing distribution

Hebert and Seaman consider the following example, where G is a discrete distribution. To facilitate exposition, we refer to Λ, the random hazard rate, as a random variable with distribution $G(\lambda)$. Let X be an exponential process with hazard rate Λ, where $P(\Lambda = \lambda_1) = p$ and $P(\Lambda = \lambda_2) = 1 - p$. It follows that

$$\mathcal{L}_t\{G(u)\} = \int_0^\infty \exp(-\lambda t)\, dG(\lambda)$$
$$= p\exp(-\lambda_1 t) + (1-p)\exp(-\lambda_2 t),$$

yielding,

$$E_\Lambda\{\Lambda^{-k} H_\Lambda(t)\} = \Big\{ p\lambda_1^{-k}\exp(-\lambda_1 t)$$
$$+ (1-p)\lambda_2^{-k}\exp(-\lambda_2 t)\Big\}$$
$$\div \Big\{ p\exp(-\lambda_1 t)$$
$$+ (1-p)\exp(-\lambda_2 t)\Big\}.$$

From expression (1.4), one obtains

$$\sigma^2(t) - \sigma^2 = p(1-p)\left(\lambda_1^{-1} - \lambda_2^{-1}\right)^2$$
$$\cdot\Big\{\big\{\exp[-t(\lambda_1 + \lambda_2)]$$
$$\div [p\exp(-\lambda_1 t) \qquad (2.1)$$
$$+ (1-p)\exp(-\lambda_2 t)]^2\big\} - 1\Big\}.$$

To illustrate the dependence of T on p, Hebert and Seaman consider the mixture with $\lambda_1 = 2$ and $\lambda_2 = 4$. Setting expression (2.1) equal to zero and using Maple, we find that the positive root of the equation is $t = \ln\left(\frac{1-p}{p}\right)$. When $p < 0.5$, this root is positive. Thus, the variance inflation boundary is given by $T = \max\left\{0, \ln\left(\frac{1-p}{p}\right)\right\}$. The actual change in variance, $\sigma^2(t) - \sigma^2$, is plotted as a function of the parameters t and p in Figure 2.1.

Using Mullooly's algorithm, one obtains a lower bound for T of approximately 0.03, as compared to the actual value $T = \ln(3) \approx 1.099$. As $p \to 0.5$, the root of (2.1)

approaches zero. Since Mullooly's lower bound remains positive, the bound improves as p nears 0.5. Likewise, the bound becomes more conservative as p nears zero.

Using Maple, one may generalize this example by considering the case $2\lambda_1 = \lambda_2$. We find in this case that

$$T = \max\left\{0, \frac{2\ln\left(\frac{1-p}{p}\right)}{\lambda_1}\right\}. \qquad (2.2)$$

When $\lambda_1 = 2$, we obtain the result discussed by Hebert and Seaman. As a more applicable generalization of this model, we offer the following result. The proof of this result is found in a Maple session in Appendix A.

Theorem: Let X be a two-point, mixed exponential process with mixing distribution $P(\Lambda = \lambda_1) = p$ and $P(\Lambda = k\lambda_1) = 1 - p$. Then the variance inflation boundary is given by

$$T = \max\left\{0, \frac{2\ln\left(\frac{1-p}{p}\right)}{\lambda_1(k-1)}\right\}. \qquad (2.3)$$

One may note that for $k = 2$, we obtain expression (2.2). The value of (2.3) is that it represents *all* possible two-point mixtures. If λ_1 and λ_2 are any points in $(0, \infty)$, one may determine the variance inflation boundary by taking $k = \frac{\lambda_2}{\lambda_1}$ and using expression (2.3). In such cases, the sensitivity of the T with respect to the mixing probability, p, can be represented graphically, as in Figure 2.1.

3. A composite mixing distribution

Properties of the mixture (1.2) when Λ has a

gamma distribution are considered by McNolty (1964), Harris and Singpurwalla (1968), and McNolty, Doyle, and Hansen (1980). Characteristics of the variance of this model under truncation are considered by Hebert and Seaman. As a generalization of the gamma model, Hebert and Seaman consider a gamma mixture with a point mass at an arbitrary value λ_0. We address that example as well in the present section. All derivations which appear in the Hebert and Seaman example were calculated using Maple. In this paper, we discuss these calculations briefly and use Maple graphics to examine some further characteristics of this model under truncation.

Let X represent the lifetime of a component which operates in a fixed environment with probability p and a random environment otherwise. When operating in the fixed environment, X has an exponential distribution with hazard rate $h(x) = \lambda_0$. In the random environment, the process is still exponential, but the hazard rate varies with distribution

$$g(\lambda) = \frac{\beta^\alpha}{k^* \Gamma(\alpha)} \lambda^{\alpha - 1} \exp\left[-\beta(\lambda - \lambda_0)\right], \lambda > \lambda_0,$$

where $k^* = \sum_{i=0}^{\alpha-1} \frac{(\lambda_0 \beta)^i}{i!}$. Thus, the density of X is a mixture of exponential densities with mixing distribution

$$G(\lambda) = pJ_{\lambda_0}(\lambda) + \frac{(1-p)}{k^*}$$

$$\cdot \sum_{i=0}^{\alpha-1} \frac{(\lambda_0 \beta)^i}{i!} \frac{\gamma(\alpha-i, \beta(\lambda-\lambda_0))}{\Gamma(\alpha-i)},$$

where $\lambda_0 > \lambda$, $J_{\lambda_0}(\lambda)$ is a unit step function at $\lambda = \lambda_0$ and $\gamma(\alpha - i, \beta(\lambda - \lambda_0))$ is the incomplete gamma function with parameter

$\alpha - i$ evaluated at $\beta(\lambda - \lambda_0)$. Direct calculation provides

$$\mathcal{L}_t\{G(t)\} = \exp(-\lambda_0 t)\left\{p + \frac{(1-p)}{k^*}\right.$$
$$\left. \cdot \left(\frac{\beta}{\beta+t}\right)^\alpha \sum_{i=0}^{\alpha-1} \frac{(\lambda_0(\beta+t))^i}{i!}\right\}$$

and (3.1)

$$E_\Lambda\left\{\Lambda^{-k} H_\Lambda(t)\right\} = \left\{p\lambda_0^{-k} + \frac{(1-p)}{k^*}\right.$$
$$\cdot \sum_{i=0}^{\alpha-1} \frac{(\lambda_0\beta)^i \beta^{\alpha-i}}{i!\,\Gamma(\alpha-i)}$$
$$\cdot \int_0^\infty \frac{y^\alpha e^{-y(\beta+t)}}{y^{i-1}(y+\lambda_0)^k}\,dy\right\}$$
$$\div \left\{p + \frac{(1-p)}{k^*}\right.$$
$$\left. \cdot \left(\frac{\beta}{\beta+t}\right)^\alpha \sum_{i=0}^{\alpha-1} \frac{(\lambda_0(\beta+t))^i}{i!}\right\}.$$

As in McNolty, Doyle, and Hansen (1980), we must have $\alpha > k$ for these moments to exist. As an illustration of the characteristics of the difference $\sigma^2(t) - \sigma^2$, Hebert and Seaman consider this model with $\beta = 1$ and $\alpha = 3$. Using Maple one may show that (3.1) and (1.4) yield expression (3.2)

$$\sigma^2(t) - \sigma^2 = \left\{2p\lambda_0^{-2}(1-\xi(t)) + 2\Psi\left[(t+1)^{-1}\right.\right.$$
$$\left. - \xi(t)\right]\}\xi^{-1}(t) + \left\{p^2\lambda_0^{-2}(\xi(t)^2\right.$$
$$-1) + \Psi^2[\xi^2(t)(1+\lambda_0)^2$$
$$-((t+1)^{-2} + \lambda_0(t+1)^{-1})^2]$$
$$+ 2p\lambda_0^{-1}\Psi[\xi^2(t)(1+\lambda_0) - ((t$$
$$+1)^{-2} + \lambda_0(t+1)^{-2})]\}\xi^{-2}(t)$$

where $\Psi = \frac{1-p}{\lambda_0^2 + 2\lambda_0 + 2}$ and $\xi(t) = p + \Psi[\lambda_0^2(t+1)^{-1} + 2\lambda_0(t+1)^{-2} + 2(t+1)^{-2}]$. One

70

may easily verify that $\xi(0) = 1$ and $\lim_{t \to \infty} \xi(t) = p$.

Using these results, we evaluate (3.2) at 0 and taking the limit as $t \to \infty$, we find $\sigma^2(0) - \sigma^2 = 0$ and

$$\lim_{t \to \infty} \sigma^2(t) - \sigma^2 = (1-p)[2\lambda_0^3 + \lambda_0^2(5-p)$$
$$- 4\lambda_0(p-2) - 4(p-1)]$$
$$\div \lambda_0^2(\lambda_0^2 + 2\lambda_0 + 2)^2$$
$$> 0. \qquad (3.3)$$

The derivative of the difference (3.2) with respect to t is easily shown to be positive using Maple. One may also verify that the derivative is completely monotone, indicating that (3.2) approaches (3.3) monotonically. Thus, truncation of this mixture will increase variance at all $t \in (0, \infty)$, when p and λ_0 are fixed, i.e., $T = \infty$. Recall that in such cases Mullooly's algorithm is inapplicable, since it assumes that the difference attains a maximum.

In Figure 3.1, we consider the behavior of (3.2) with $p = 1/2$ for various values of λ_0 and t. Some notable features of this model are evident in this graph. As λ_0 increases, the process begins to *behave* as though it were a pure exponential process, i.e., truncation does not effect variance. One may also note that the contour lines on the plot reaffirm that the inflation in variance is bounded as a function of t when $\lambda_0 > 0$ is fixed.

In the *pure gamma process*, $\lambda_0 = 0$, the variance increase is unbounded as a function of t, see Hebert and Seaman. The point mass included in the present model dampens the variance increase observed in the pure gamma

process. The sensitivity of this increase with respect to the values, λ_0 and t, is evident in Figure 3.1. A similar behavior is evident when λ_0 is fixed. In Figure 3.2, we plot the difference (3.2) as a function of p and t with $\lambda_0 = 1/2$. Again, we see that for $p = 0$ and $p = 1$, we have the *pure gamma* behavior noted by Hebert and Seaman and the pure exponential behavior, respectively. This is consistent with the form of the model, since $p = 1$ would indicate the the process does not operate in a random environment at any time and with $p = 0$, the process remains in the random environment and possesses a gamma-distributed hazard rate.

Appendix A
Proof of the Theorem of Section 2

First define sd to be the difference in the variance generated by truncation.

```
> sd:=t->p*(1-p)*(lambda[1]^(-1)
  -lambda[2]^(-1))^2*(exp(-t*(
  lambda[1]+lambda[2]))/(p*exp(
  -lambda[1]*t)+(1-p)*exp(-lambda
  [2]*t))^2 -1);
```

$$sd := t \mapsto p\,(1-p) \left(\frac{1}{\lambda_{[1]}} - \frac{1}{\lambda_{[2]}} \right)^2$$
$$\left(\frac{e^{-t(\lambda_{[1]} + \lambda_{[2]})}}{pe^{-\lambda_{[1]}t} + (1-p)e^{-\lambda_{[2]}t}} - 1 \right)$$

Set:

```
> lambda[1] := lambda;
  lambda[2] := k*lambda[1];
  'sd(t) =',sd(t);
```

$$\lambda_{[1]} := \lambda$$
$$\lambda_{[2]} := k\lambda$$
$$sd(t) =, p\,(1-p) \left(\frac{1}{\lambda} - \frac{1}{k\lambda} \right)^2$$

71

$$\left(\frac{e^{-t(\lambda + k\lambda)}}{pe^{-\lambda t} + (1-p)e^{-k\lambda t}} - 1 \right)$$

We look at the numerator then perform a series of transformations to facilitate finding roots.

```
> numer(sd(t)) = 0:
  eq_1 := combine( ", exp);
```

$$eq\_1 := p(-1 + p)(k-1)^2(-e^{-t(\lambda + k\lambda)}$$
$$+ p^2 e^{-2\lambda t} + 2pe^{-\lambda t - k\lambda t}$$
$$- 2p^2 e^{-\lambda t - k\lambda t} + e^{-2k\lambda t}$$
$$- 2pe^{-2k\lambda t} + p^2 e^{-2k\lambda t}) = 0$$

```
> subs( t=ln(T), eq_1 / (p*(-1+p)*(k-1)^2) ):
  eq_2 := simplify( ", exp );
```

$$eq\_2 := -T^{(-\lambda(1+k))} + p^2\, T^{(-2\lambda)}$$
$$+ 2p T^{(-\lambda(1+k))} - 2p^2 T^{(-\lambda(1+k))}$$
$$+ T^{(-2k\lambda)} - 2p T^{(-2k\lambda)}$$
$$+ p^2 T^{(-2k\lambda)} = 0$$

```
> subs( T=z^(1/lambda), eq_2 ):
  eq_3 := simplify( ", power );
```

$$eq\_3 := -z^{(-1-k)} + \frac{p^2}{z^2} + 2pz^{(-1-k)}$$
$$- 2p^2 z^{(-1-k)} + z^{(-2k)} - 2p^2 z^{(-2k)}$$
$$+ p^2 z^{(-2k)} = 0$$

```
> eq_4 := simplify( eq_3*( z^(2*k) ),power );
```

$$eq\_4 := -z^{(k-1)} + z^{(2k-2)}p^2 + 2z^{(k-1)}p$$
$$- 2z^{(k-1)}p^2 + 1 - 2p + p^2 = 0$$

```
> simplify(subs(z=Z^(1/(k-1)), eq_4 ), power):
  eq_5 := collect( ", Z );
```

$$eq\_5 := Z^2 p^2 + (-2p^2 - 1 + 2p)Z + 1$$
$$- 2p + p^2 = 0$$

We have arrived at an expression quadratic in Z which is simple to solve.

```
> soln_Z := factor([solve(eq_5,{Z})]);
```

$$soln\_Z := \left[\{Z = 1\}, \left\{ Z = \frac{(-1+p)^2}{p^2} \right\} \right]$$

Now back-substitutions will finish the derivation.

```
> soln_Z := subs( -1+p=1-p, soln_Z ):
  soln_z := subs( Z=z^(k-1), soln_Z ):
  soln_T := subs( z=T^lambda, soln_z ):
  soln_t := subs( T=exp(t), soln_T );
```

$$soln\_t := \left[\left\{ \left((e^t)^\lambda \right)^{(k-1)} = 1 \right\}, \right.$$
$$\left[\left\{ \left((e^t)^\lambda \right)^{(k-1)} = \frac{(1-p)^2}{p^2} \right\} \right]$$

```
> op(map(solve,soln_t,{t})):op("[1]);
  op(""[2]);
```

$$t = \ln\left(\left(\left(-\frac{-1+2p-p^2}{p^2} \right)^{\left(\frac{1}{k-1}\right)} \right)^{\left(\frac{1}{\lambda}\right)} \right)$$

```
> t_s:=[t=0,t=2/lambda/(k-1)*ln((1-p)/p)];
```

$$t\_s := \left[t = 0, t = 2\frac{\ln\left(\frac{1-p}{p}\right)}{\lambda(k-1)} \right]$$

```
>
```

QED

Bibliography

Barlow, R. E. and F. Proschan (1975), *Statistical Theory of Reliability and Life Testing: Probability Models* (Holt, Rinehart, and Winston, New York).

Harris, C. M. and Singpurwalla, N. (1968), Life Distributions Derived from Stochastic Hazard Functions, *IEEE Trans. on Reliability*, **R-17**, 70-79.

Hebert, J. L. and Seaman, J. W. Jr. (1994), The variance of a truncated mixed exponential process, *Journal of Applied Probability*, **31**, No. 1 (In Press).

Jewell, N. P. (1982), Mixtures of Exponential Distributions, *The Annals of Statistics* **10**, 479-484.

McNolty, F. (1964), Reliability Density

Functions when the Failure Rate is Randomly Distributed, *Sankhyā*, **A-26**, 287-292.

McNolty, F. , Doyle, J. and Hansen, E. (1980), Properties of the Mixed Exponential Failure Process, *Technometrics*, **22**, 555-565.

Mullooly, J. P. (1988), The Variance of Left-Truncated Continuous Nonnegative Distributions, *The American Statistician* **42**, 208-210.

Biographies

William Bauldry received his PhD from Ohio State. His thesis and subsequent research was in Approximation Theory. Currently, he is co-principle investigator of an NSF consortium building a Maple laboratory at Winston-Salem State University. Present research interests include graph coverings and moments of distributions. Currently, he is an Associate Professor in the Department of Mathematical Sciences at Appalachian State University, Boone, NC 28608.

Jaimie L. Hebert received his PhD in Statistics from the University of Southwestern Louisiana in May 1990. His research interests include characteristics of exponential mixtures, estimation with truncated data, and estimation of location parameters in a random environment. Currently, he is an Assistant Professor in the Department of Mathematical Sciences at Appalachian State University, Boone, NC 28608.

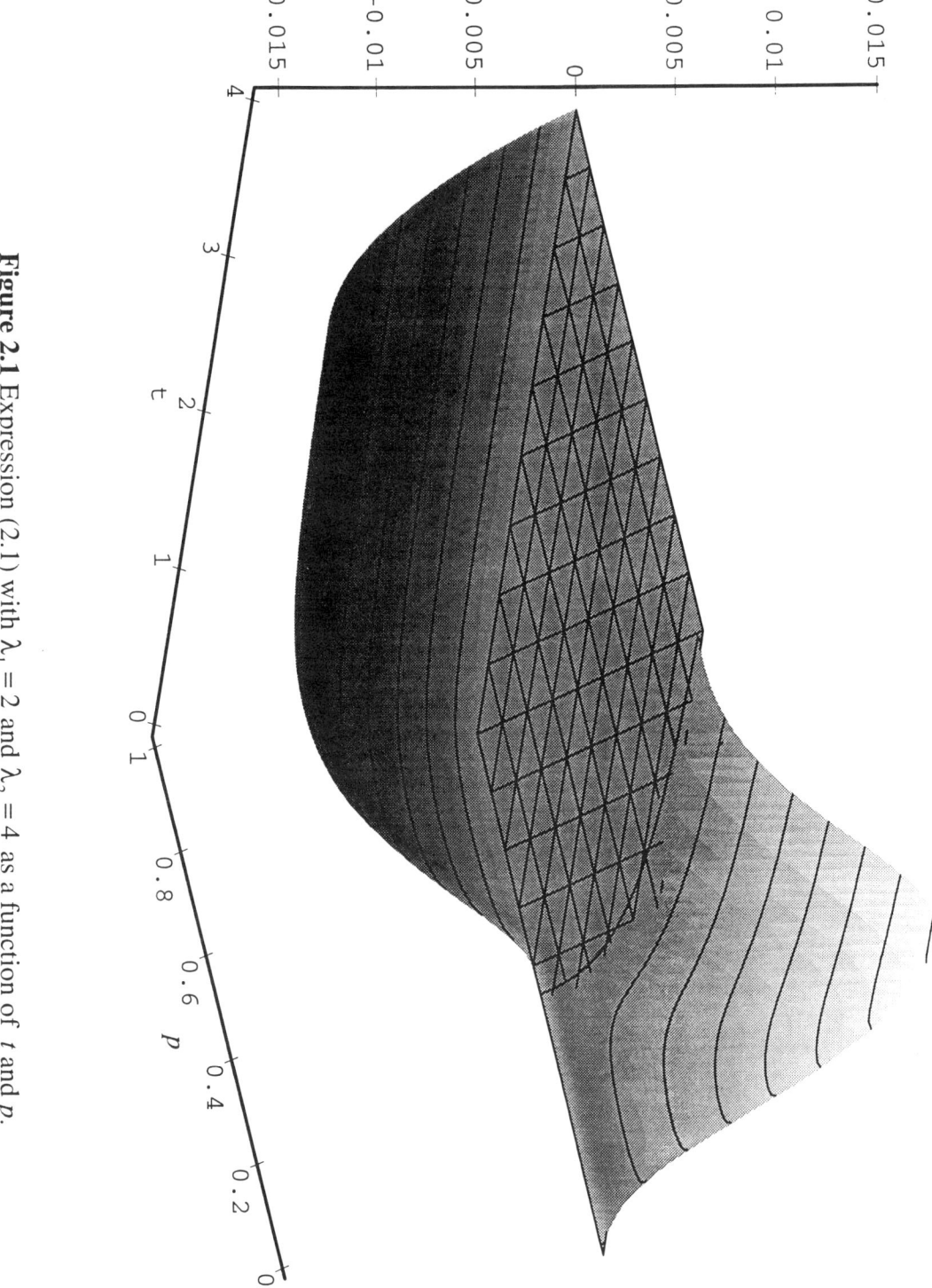

Figure 2.1 Expression (2.1) with $\lambda_1 = 2$ and $\lambda_2 = 4$ as a function of t and p.

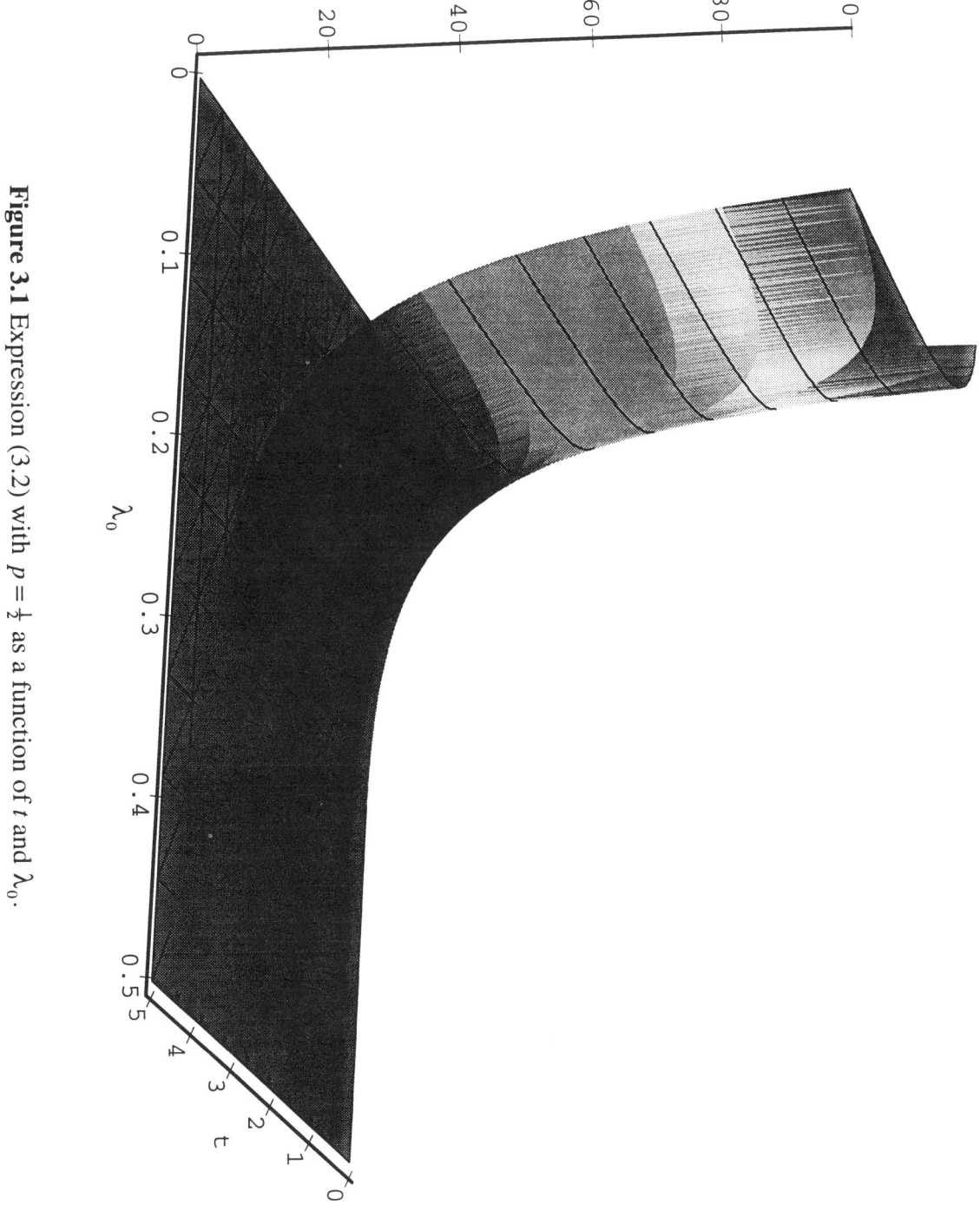

Figure 3.1 Expression (3.2) with $p = \frac{1}{2}$ as a function of t and λ_o.

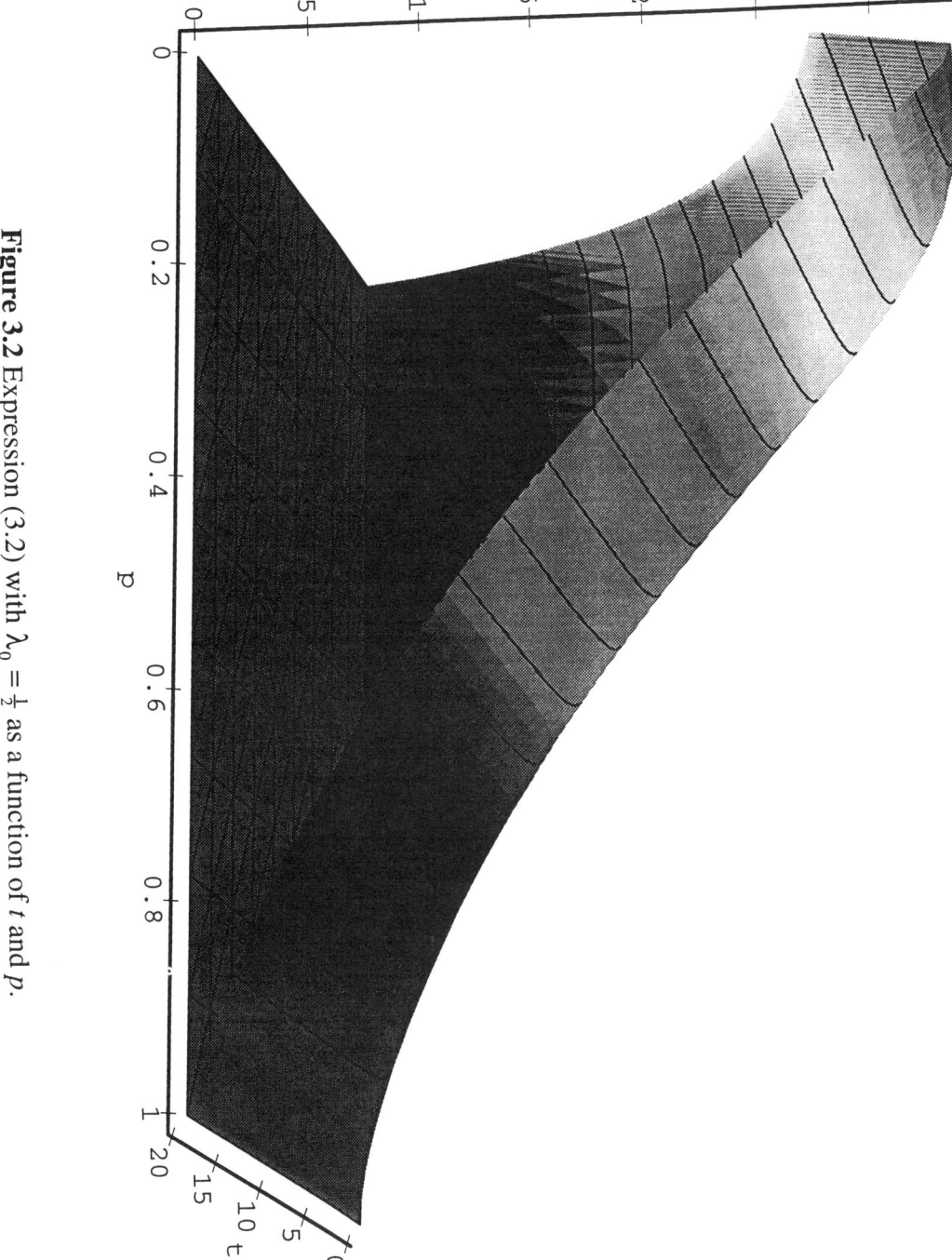

Figure 3.2 Expression (3.2) with $\lambda_0 = \frac{1}{2}$ as a function of t and p.

WORKING WITH LARGE MATRICES IN MAPLE

Reid Pinchback
Academic Computing Services, Massachusetts Institute of Technology,
Cambridge MA, USA

Introduction

Many engineering and scientific problems can be solved by linear methods, but non-trivial examples of these problems may require large amounts of memory to represent and even larger amounts of computing time to solve. As more degrees of freedom, n, are added to a problem, the growth in memory and processing time can be $O(n^2)$ and $O(n^3)$, respectively.

This is a situation that experienced faculty and research staff are generally well acquainted with, and they have learned to effectively match the problem size and choice of algorithms with the available computing power. On the other hand, students — and particularly graduate students — tend to underestimate the computational demands of problems they attempt to solve; the situation deteriorates further when the algorithms they use are chosen for their simplicity instead of their computational effectiveness. In a multiuser timesharing facility, the resulting long and large jobs can present a strain on the available resources.

History

A situation where a computation is very big or takes very long is undesirable, because generally the student does not have access to more powerful equipment to alleviate the performance problems. At MIT we were experiencing a small, but continual, stream of requests from students

Mathematical Computation with Maple V:
Ideas and Applications
Tom Lee, Editor

that needed to run uninterrupted computations for one or two days. Since the single-user workstations that form the bulk of our resources are only available to students for short periods of time, we worked to provide a batch computing facility that could handle these special cases.

The resource provided for batch processing was a very fast VAX/Ultrix machine — the VAX 9000 — configured as a multiuser timeshared facility. When students began running jobs on it, the result was a fast machine that periodically thrashed to the point of being unusable. The hunt was on to find the source of the problem.

It turned out that a small number of users were simultaneously running a few large jobs, often to solve large linear problems. The effect was increased swapping of virtual memory pages, which increased system overhead, which in turn resulted in a decrease in available computing time for users, which resulted in more frequent task switching, and so on through the vicious cycle. We learned that a situation that was merely undesireable for a single-user workstation became dramatically worse in a time-shared environment, primarily because these users were forcing the virtual memory system to manage their data for them instead of managing it themselves.

To set the problem in a clearer light, the virtual memory system creates an address space larger than the available RAM, and swaps memory pages to and from disk as they are needed. This works well when the rate of page swapping is small relative to the rate of process task switches. When some memory pages are accessed more frequently than others, or when the amount of virtual memory currently being exercised is small, page swapping will be infrequent.

Consider a simple situation where two users initiate processes to manipulate large arrays, and assume that these processes are similar in size to the physical memory (RAM) of the computer. When the two user processes start processing the arrays, operations like array addition will have the effect of refreshing the access time on all of the pages containing the array data. For the current process, this will force its data pages to be swapped into memory, and most other pages to be swapped out. When the next task switch happens, the new process will be missing its data pages, so they must be swapped in after other pages are swapped out to make room. Soon, the system is spending most of its time on swapping memory pages instead of being able to allocate CPU time for user process calculations.

Since part of our end-user support philosophy at MIT is to help users to become self-sufficient, we began considering ways in which students could approach solving large problems differently. The ultimate aim was to document methods and provide on-line code samples that would help them to solve large linear problems more effectively, and towards that end we considered two issues:

1) Reducing the memory needed for storing array data.
2) Designing long-running programs that could be recovered and restarted after interruption.

This paper describes an attempt to address the first of these two issues.

Method

One way to reduce the memory demands of large arrays is to partition those arrays into smaller sections for processing, only loading a few of those sections from disk into virtual memory as they are needed.

The resulting computation has more apparent overhead because it must manage its own data, but because of the reduced demands on virtual memory the result is a computation that can run smaller and faster in a time-shared environment.

Maple provides a good prototyping environment for approaching this problem. The challenge was to architect an implementation that would be transparent to existing Maple routines, thus avoiding the extensive work otherwise needed to re-implement packages like linalg. The indexfcn facility, when combined with a pre-processor to the array function, makes it possible to change the semantics of array creation and access without requiring changes to existing libraries of matrix routines.

Implementation

The partitioning method used here is a physical scheme, as opposed to a mathematical one. Since Maple routines like 'linalg/add' are not aware of the partitioning, they do not attempt to manipulate the arrays on a partition-by-partition basis. Instead, a partition cache is used to increase the proportion of hits in accessing partitioned arrays.

The partitions themselves are a collection of flat (one-dimensional) unpartitioned arrays. The original array indicies are mapped onto the partitions. Such a mapping is potentially different for each partitioned array, as it must reflect the width of the ranges and the number of dimensions.

In order to work with partitioned arrays, we needed three particular facilities or behaviours:

1) A way to calculate with partitioned arrays.
2) A way to create partitioned arrays.
3) A way to manage the partitions.

The first two issues will be explored in depth, while the third will only be touched on briefly.

```
# This is the core routine of the partitioning indexfcn.
'index/partitioned' := proc(A,index,isLHS,tbl)
  local i,range,position,partition,pindex,phandle,p;
  # calculate position, the flattened index into the partitions
  position := 0;
  for i from nops(index) by -1 to 1 do
    range := tbl['_ranges'][i];
    position := 'parray/rangewidth'(range)*position + index[i] - op(1,range);
  od;
  partition := iquo(position, tbl['_Partitionsize']);
  pindex    := irem(position, tbl['_Partitionsize']);
  # phandle[partition] will be the partition number, p
  phandle := invokePHandle(tbl['_phandle']);
  # check to see if the partition exists yet
  if not assigned(phandle[partition]) then
    if isLHS then
      phandle[partition] := generatePartition(tbl['_Partitionsize']);
    else
      op(index);
      RETURN('A["]');
    fi;
  fi;
  p := phandle[partition];
  # check to see if the particular element, pindex, of the partition exists
  if isLHS then
    pindex;
    partitionCache(p);
    RETURN('"[""]');
  else
    if not assigned(partitionCache(p)[pindex]) then
      op(index);
      RETURN('A["]');
    else
      RETURN(partitionCache(p)[pindex]);
    fi;
  fi;
end: # proc index/partitioned

# This routine determines the width of an index range
'parray/rangewidth' := proc(r:indexrange)
  RETURN(op(2,r) - op(1,r) + 1);
end: # proc parray/rangewidth
```

Listing 1

Calculation

To use partitioned arrays in calculations, we need to be able to store data to — and retrieve data from — the array. In order to do this, we create an *indexfcn*. When a new indexfcn is implemented correctly, it will allow assignment to an array element, retrieval of an array element, and it will cope with unassigned elements properly.

All we need is an indexfcn that understands how the arrays are partitioned.

Each partitioned array is given its own unique indexfcn which has a table containing some basic information used to locate the partitions belonging to that array. Each of these indexfcn's call a common routine, 'index/partitioned', to do the actual work (see listing 1).

The `'index/partitioned'` routine, together with `'parray/rangewidth'`, maps the array index into a corresponding partition number and an index into that partition. When referenced to the partition number array (`phandle`) for the particular array, this determines where the desired element would be stored.

If an assignment is being attempted ("`if isLHS`"), and the corresponding partition has not yet been created, it gets created here. If a retrieval is being attempted, the appropriate element — if it exists — is returned.

Creation

Creation of partitioned arrays is not a particularly difficult issue if arrays are only created intentionally by the user. Unfortunately, packages like `linalg` internally create non-partitioned arrays during the course of computation, so we needed a way to make those routines change their default array creation semantics. Since Maple doesn't have any kind of class inheritance to make use of, if we are to avoid re-writing existing library code the only workable solution is to change the `array` function itself.

There is one complication we must address before proceeding further. Since `array` is a built-in procedure that we need in order to create the actual array data structures, in replacing it we still must have a way to invoke it. Installation of the new facility (called `parray`) must be handled carefully (see listing 2).

The changes to `array` implemented in

```
if not assigned('parray/array') then
  # retain access to built-in 'array'
  'parray/array' := op(array):
  # change to the new functionality
  array := parray:
  # to handle type renaming confusion
  'type/parray' := 'array':
fi:
```

Listing 2

`parray` must be transparent if existing code is to continue working reliably. Arguments must appear to be processed in the same way that `array` normally processes them, and the results returned must correspond to those normally returned.

Proper argument handling is possible, but serves to show a rather unexpected limitation of the Maple type expression facilities. A valid argument list for `array` consists of — in any order — an optional index function, an optional list of initialization elements, and an optional contiguous block of an arbitrary number of index ranges. This corresponds to the type grammar shown in figure 1.

No Maple type expression can readily describe this kind of argument list. Maple type expressions for structured types (like `list(integer)`) can apparently only describe types corresponding to:

- A structured type with a known number of components with heterogeneous types.
- A hierarchical type with components all of a single type.

```
⟨arglist⟩      ::=  [ ⟨indexfcn-arg⟩ , ⟨range-args⟩ , ⟨initialization-arg⟩ ]
                  | [ ⟨initialization-arg⟩ ⟨range-args⟩ ⟨indexfcn-arg⟩ ]
⟨indexfcn-arg⟩      ::=  string | procedure | λ
⟨initialization-arg⟩ ::=  list(anything) | λ
⟨range-args⟩      ::=  ⟨range-arg⟩ , ⟨range-args⟩ | λ
⟨range-arg⟩       ::=  range(integer)
```

Figure 1

This means that type expressions like:

[integer, float, integer, float]

can describe the type of objects like:

[1, 2.0, 2, 3.0]

but not arbitrarily long (finite) lists like:

[1, 2.0, 2, 3.0, 3, 4.0, ...]

nor is there a type expression that describes anything remotely like:

[true, 1.0, 2.0, 3.0, ...].

There is no Maple type-expression equivalent of a regular expression or a head-versus-tail list expression syntax, so there is no convenient way to express hierarchical types like list((integer,float)) or have types described by regular expressions such as [integer, float*, integer].

If a regular expression would achieve what we need, then a procedural type that implements the regular expression would do the job. Since a regular expression has an equivalent deterministic finite automaton (See [1] for proofs and general background information), a procedural type implementing the DFA can be used. Figure 2 depicts the DFA, and listing 3 shows its implementation in Maple.

The 11-state minimal DFA implemented here is just a combination of three simple recognizers corresponding to each of the three kinds of valid arguments. The procedural type for indexrange performs a litte extra work to verify that each range a .. b satisfies the constraint a £ b.

Once we know we have a valid argument list, we can go on with processing the arguments and creating partitioned arrays (see listing 4).

Any array containing more than a default number of elements, Partitionsize, will be partitioned, and the current value of Partitionsize is cached with the array.

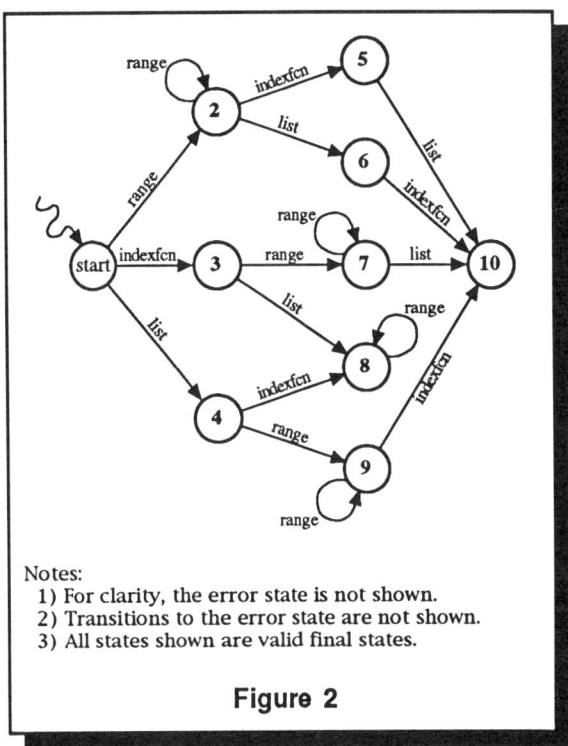

Notes:
1) For clarity, the error state is not shown.
2) Transitions to the error state are not shown.
3) All states shown are valid final states.

Figure 2

This allows the default to change for future array creations without invalidating the partitioning assumptions of existing arrays.

Any indexfcn other than partitioned will be processed by the built-in array functionality, so there is no such thing as a diagonal or symmetric array that is also partitioned. If necessary, ranges will be set and initialization elements processed by the undocumented Maple routines 'table/initbds' and 'table/initvals', respectively.

Partition Management

Partition management is primarily an issue of creating a scheme for deciding when to move data to and from disk, and storing the data in such a way as to defeat unwanted garbage collection. A crude prototypical partition manager was created for testing the creation of and calculation with partitioned arrays (see listing 5). A detailed implementation that actually moves the data to and from disk has not been developed at this time, but in the future will be bundled together with developing

```
# This routine type-checks to see that the argument list passed to 'array'
# is well-formed.  To do this it uses a 11-state deterministic finite
# automaton to scan the argument list for:
#   1) at most one list of initialization elements
#   2) at most one contiguous block of valid range arguments
#   3) at most one index function, either a string (a name) or a procedure
# and these 3 groups of arguments can appear in any order, and no
# other kinds of arguments can be present.
'type/parrayargs' := proc(arglist)
  local state, arg;
  state := 'start';
  for arg in arglist do
    state := procname('DFA')[(state,`parray/typesymbol`(arg))];
  od;
  RETURN(evalb(state <> 'error'));
end: # proc type/parrayargs

'type/parrayargs'('DFA') := table([
  seq(('start',i)=op(i,[3,       2,       4,       'error']),i=1..4),
  seq((2,i)       =op(i,[5,       2,       6,       'error']),i=1..4),
  seq((3,i)       =op(i,['error',7,       8,       'error']),i=1..4),
  seq((4,i)       =op(i,[8,       9,       'error','error']),i=1..4),
  seq((5,i)       =op(i,['error','error',10,      'error']),i=1..4),
  seq((6,i)       =op(i,[10,      'error','error','error']),i=1..4),
  seq((7,i)       =op(i,['error',7,       10,      'error']),i=1..4),
  seq((8,i)       =op(i,['error',8,       'error','error']),i=1..4),
  seq((9,i)       =op(i,[10,      9,       'error','error']),i=1..4),
  seq((10,i)      =op(i,['error','error','error','error']),i=1..4),
  seq(('error',i)=op(i,['error','error','error','error']),i=1..4)
]): # table type/parrayargs(DFA)

# Map the three kinds of argument types into the integers {1,2,3}:
#   index function -> 1
#   index range    -> 2
#   list           -> 3
#   anything else  -> 4
# This is used to simplify examination of the type of arguments passed
# to the 'array' procedure.
'parray/typesymbol' := proc(arg:anything)
  if type(arg, 'string') or type(arg, 'procedure') then
    RETURN(1);
  elif type(arg, 'indexrange') then
    RETURN(2);
  elif type(arg, 'list') then
    RETURN(3);
  else
    RETURN(4);
  fi;
end: # proc parray/typesymbol

# This type-checks for a valid array index range.
# The argument must be an integer range a .. b where a <= b.
'type/indexrange' := proc(arg)
  type(arg, 'range(integer)') and evalb(op(1,arg) <= op(2,arg));
end: # proc type/indexrange
```

Listing 3

```
if not assigned(Partitionsize) then
  Partitionsize := 10:
fi:

parray := proc()
  local indexfn,ranges,elements,volume,partitionable,A,partitions,tbl,iproc;
  if not type([args], cat(procname,'ar','gs')) then
    ERROR('Error, invalid parameters for creation of table or array');
  fi;
  # read routines needed from the library:
  readlib('table/initbds');
  readlib('table/initvals');
  # Since the argument list is well formed, we can safely separate the
  # three kinds of arguments for further (and easier) examination.
  indexfn  := op(select(x->evalb('parray/typesymbol'(x)=1), [args]));
  ranges   := select(x->evalb('parray/typesymbol'(x)=2), [args]);
  elements := op(select(x->evalb('parray/typesymbol'(x)=3), [args]));
  # Determine what the ranges should be for the list of elements
  if (ranges = []) and (elements <> NULL) then
    ranges := ['table/initbds'(elements)];
  fi;
  # Check to see if the total number of elements in the array
  # (the "volume" of the array) warrants partitioning.
  volume := convert(map('parray/rangewidth', ranges), '*');
  partitionable := evalb(volume >= Partitionsize);
  # Now, see if we need to specify a partitioned index function
  if indexfn = NULL and partitionable then
    indexfn := 'partitioned';
  fi;
  # Call the appropriate routine to create and initialize the array data structure.
  if (indexfn <> 'partitioned') or not partitionable then
    A := 'parray/array'(indexfn,op(ranges),elements);
  else
    # ok, time to finally do some partitioning
    partitions := 1 + iquo(volume-1,Partitionsize);
    # store partition size, partitions needed, ranges,
    # and partition table handle in the index function
    tbl := table(['_Partitionsize'=Partitionsize,
                  '_partitions'=partitions,
                  '_ranges'=ranges,
                  '_phandle'=generatePHandle(partitions)]);
    iproc := proc(A,index,isLHS)
                'index/partitioned'(A,index,isLHS,'TBL')
             end;
    iproc := subs('TBL'=op(tbl),op(iproc));
    A := subsop(1=op(iproc),'parray/array'(op(ranges)));
    # Finally, initialize the elements of the partitioned array
    if elements <> NULL then
      'table/initvals'(op(A),elements);
    fi;
  fi;
  op(A);
end: # proc array
```

Listing 4

```
if not assigned(phandles) then
  phandles := -1:
fi:

generatePHandle := proc(parts)
  phandles := phandles+1;
  phandledata[phandles] := 'parray/array'(0 .. parts-1);
  phandles;
end: # proc generatePHandle

partitionCache := proc(pname)
  # very simplistic for now, just keeps accumulating partitions
  if not type(pname,name) then
    ERROR('invalid name argument, ',pname);
  fi;
  procname(pname) := pname;
end: # proc partitionCache

if not assigned(partsGenerated) then
  partsGenerated := 0:
fi:

generatePartition := proc(partSize)
  local parts;
  if not type(partSize,posint) then
    ERROR('invalid partition size, ',partSize);
  fi;
  parts := partsGenerated;
    partsGenerated := partsGenerated+1;
  '_p'.parts := 'parray/array'(0 .. partSize-1);
  partitionCache('_p'.parts);
end: # proc generatePartition
```

Listing 5

mechanisms for recovering long-running interrupted jobs.

Array Semantics

There is more than one category of semantic activity or interpretation happening when we consider array access as it is implemented by the indexfcn facility. These categories can be loosely divided into:

1) Storage/retrieval semantics — When we store a value into an element of an array, just where does that value really get put? These semantic actions implement the storage/retrieval of the unique data content of an array.

2) Functional semantics — Given only the index of an element in an array, how can its value be determined independent of storage? These semantic actions implement or constrain the mathematically interesting properties of an array, such as symmetry or diagonality.

3) Value semantics — Given the index of an element of an array, determine its value. These semantics determine how to intepret or prioritize the combined results of the storage/retrieval semantics and the functional semantics.

Unfortunately, these three semantic intepretations are implemented in a rather muddled combination within Maple indexfcns. Here are two examples to clarify the categories of semantic activity:

1) Consider the `identity` indexfcn. With this function, no elements need storage in the array, so the only storage/retrieval semantics are to generate an error whenever an assignment is attempted. The functional semantics determine that the diagonal elements have value 1, and all off-diagonal elements have value 0. The value semantics rely totally upon the functional semantics to determine the ultimate value of the referenced element.

2) Consider the `sparse` indexfcn. Here, the only storage space needed is for the elements actually stored, so the storage/retrieval semantics involve locating and storing/retrieving the elements from the appropriate data structure. The functional semantics can be viewed as giving all elements the value 0. The value semantics return the retrieved value if such exists, otherwise return the result of the functional semantics.

Having all three kinds of semantics works well enough for these two examples, but there are some unfortunate limitations to the approach used by Maple:

- Non-trivial indexfcn routines are difficult to implement and debug, in part because you have to figure out which pieces of semantic activity to deal with at various points in the code. Getting all the cases covered properly can be tricky, and is almost guaranteed to exercise several obscure bugs in Maple itself.

- There is no way to generate new indexfcn combinations from the desireable properties of existing indexfcns. If we have `sparse` and `symmetric` then the only way we can have `sparse_symmetric` is by going to the further effort of creating it. If we have `partitioned` and `diagonal`, then the only way to have `partitioned_diagonal` is again to implement it from scratch.

- Existing library code has no awareness of the new indexfcns that have been created, and thus no way to benefit from them. What is needed is a better mechanism for specifying indexfcn properties. This would allow users to extend library functionality without having to re-implement it and without resorting to 'back door' approaches like re-directing calls to the `array` function.

Limitations

There are several important limitations to the current attempt to prototype partitioned arrays which must be acknowledged, and addressed in the future.

- Partitioning criteria: at present, arrays are partitioned according to the number of elements in the array, and not according to the size of the elements. For numerical applications this is not a problem since all elements are the same size, typically the size of hardware floats. For more general use in Maple, the type and size of individual elements can vary greatly.

- Lifetime of elements: types that obey last-name evaluation are not viable array element types for partitioning, since only their names are stored in the array partitions. The problem with this is that the original objects may be garbage collected while a partition is swapped out.

- Use of partitions between sessions: at present, partitions behave much like Unix swap space. A partitioned array is only guaranteed to be meaningful for the current Maple session; storing it in a file and attempting to use it in future sessions won't work since only the array data structure would be saved, not the content of the partitions. This is part of the work on recoverability that will be attempted in the future.

- Algorithm performance: the design of an algorithm will affect the performance penalities experienced when using partitioning. An algorithm that has relatively random access within an array can cause a large amount of partition-swapping activity, while algorithms with relatively concentrated hits will work effectively with a moderately small partition cache.

- Arrays are not the only large objects: only array partitioning is implemented here, but large grids and tables could equally benefit from the ability to partition. Partitioning grids might be particularly useful since it would allow users to create more complicated plot structures on machines with a small amount of RAM and swap space.

- Use of undocumented and built-in routines: while the approach described in this paper was an effective way of trying to extend Maple functionality, it is a method that is likely to be unreliable in the long run. It is handy for prototyping code, but not satisfactory for creating products for end-users. Whenever a back-door approach like this is used, it probably reflects either a weakness in the design, a weakness in Maple functionality, or both.

Conclusions

The Maple indexfcn facilities provide many opportunities for extending the existing functionality of arrays. Unfortunately, with the current implementation it is difficult for users to realize its fullest potential. This is true of other aspects of Maple extensibility, as shown by the usefulness but occasional awkwardness of type specification. Continued improvement in these aspects of the Maple language can only benefit users and developers alike.

Acknowledgements

I would like to thank my co-workers, Anne Lavin and Michal Lusztig, for a variety of background information and ruminations that provided the initial motivation for this work. I would also like to thank Naomi Schmidt for her review of the various abstracts and drafts that ultimately evolved into this paper.

References

[1] Derick Wood, *Theory of Computation*, John Wiley & Sons, New York, (1987).

Reid M. Pinchback is a Faculty Liaison with the Academic Computing Services department of MIT, responsible for maintaining Unix and Macintosh installations of Maple and other software packages for educational use. He writes introductory Maple documentation and provides support and training for end users and other MIT computer consultants. email address: *reidmp@mit.edu.*

USING MAPLE FOR ASYMPTOTIC CONVERGENCE ANALYSIS

Noah H. Rhee
Department of Mathematics, University of Missouri, Kansas City MO, USA

Introduction

With prevailing trend of parallel processing diagonalization methods for solving eigenvalue problems came into renewed interest of numerical analysts. In order to study the asymptotic convergence rate of such methods it is useful to use the qualitative analysis with O-symbol. The $O(\ )$ term in a series indicates the order of truncation of the series.

The Paardekooper method is a typical diagonalization method for solving eigenvalue problem of skew-symmetric matrices. In this paper we show how Maple's facilities can be used to construct and analyse an example that demonstrates the failure of the quadratic convergence of the Paardekooper method in the presence of nonzero eigenvalues whose multiplicities are greater than two. The example is presented and analysed mathematically in [4]. Here we focus on how to use Maple. We will see that without Maple the analysis is extremely tedious.

Paardekooper Method

It is well known that any skew-symmetric matrix A is orthogonally similar to a matrix, which is the direct sum of 2 x 2 matrices of the form

$$\begin{bmatrix} 0 & \nu \\ -\nu & 0 \end{bmatrix}$$

corresponding to pairs of complex conjugate (purely imaginary) eigenvalues $\pm i\nu$ ($\nu > 0$)

Mathematical Computation with Maple V:
Ideas and Applications
Tom Lee, Editor
©1993 Birkhäuser Boston

of A and zero matrix corresponding to zero eigenvalues of A. This direct sum is called Murnaghan form [7] of A.

The Paardekooper method [8] is a Jacobi-type method for reducing a skew-symmetric matrix to its Murnaghan form. We note that skew-symmetric eigenvalue problems appear in practice (e.g. structural mechanics) [1, 9].

Let $A = (a_{ij})$ be a skew-symmetric matrix of order n. In order to avoid inessential difficulties in the description of the method, we assume n to be even. We write

$$A = \begin{bmatrix} A_{11} & \cdots & A_{1k} \\ \vdots & \ddots & \vdots \\ A_{k1} & \cdots & A_{kk} \end{bmatrix}, \qquad (1)$$

where each A_{ij} is 2 x 2 and hence $k = n/2$. Instead of (1) we also write $A = (A_{pq})$. With the partition (1) is associated

$$S(A) = \|A - diag(A_{11}, \ldots, A_{kk})\|, \qquad (2)$$

which is called *departure from the Murnaghan form*. In (2) as well as throughout the whole paper $\|\cdot\|$ stands for the Euclidean matrix norm, that is, $\|A\|^2 = trace(A^T A)$.

One *step* of the Paardekooper method uses an orthogonal similarity transformation (cf. [8])

$$A' = \mathfrak{S}_{pq}^T A \, \mathfrak{S}_{pq}, \quad p < q,$$

to make A'_{pq} zero. The matrix $\mathfrak{S}_{pq}$ is called [2, 3] *Jacobi annihilator* of A_{pq}, and A_{pq} is called *pivot submatrix*.

The tools of describing $\mathfrak{S}_{pq}$ are plane rotations. Let $R(i, j; \phi)$ stand for the rotation matrix in the (i, j) plane by the angle ϕ, that is,

$$R(i, j; \phi) = \qquad\qquad (3)$$

$$\begin{bmatrix} 1 & \cdots & 0 & \cdots & 0 & \cdots & 0 \\ \vdots & \ddots & \vdots & & \vdots & & \vdots \\ 0 & \cdots & \cos\phi & \cdots & \sin\phi & \cdots & 0 \\ \vdots & & \vdots & \ddots & \vdots & & \vdots \\ 0 & \cdots & -\sin\phi & \cdots & \cos\phi & \cdots & 0 \\ \vdots & & \vdots & & \vdots & \ddots & \vdots \\ 0 & \cdots & 0 & \cdots & 0 & \cdots & 1 \end{bmatrix}$$

where the positions of $\cos\phi$ are (i, i) and (j, j). $R(i, j; \phi)$ is called the rotation matrix in the (i, j) plane by the angle ϕ because premultiplication by $R(i, j; \phi)^T$ amounts to a counter clockwise rotation of ϕ radians in the (i, j) coordinate plane. We will call the transformation $A \longrightarrow R(i, j; \phi)^T A R(i, j; \phi)$ a rotational transformation or a transformation involving $R(i, j; \phi)$. Note that $R(i, j; \phi)$ is orthogonal, and postmultiplication [premultiplication] by $R(i, j; \phi)$ $[R(i, j; \phi)^T]$ affects only the ith and jth columns [rows].

Using plane rotations $\Im_{pq}$, $p < q$, can be written

$$\Im_{pq} = R^1_{pq} R^2_{pq} R^3_{pq} R^4_{pq},$$

where

$$\begin{aligned}
R^1_{pq} &= R(2p-1, 2q-1; \phi_1), \\
R^2_{pq} &= R(2p, 2q; \phi_2), \\
R^3_{pq} &= R(2p-1, 2q; \phi_3), \\
R^4_{pq} &= R(2p, 2q-1; \phi_4).
\end{aligned}$$

So one step of the Paardekooper method uses four rotational transformations. The transformation involving R^1_{pq} $[R^3_{pq}]$ modifies the elements at positions $(2p-1, 2q)$, $(2p, 2q-1)$ $[(2p-1, 2q-1), (2p, 2q)]$ in such a way that the transformation involving R^2_{pq} $[R^4_{pq}]$ can annihilate them. The skew-symmetry property ensures that zeros obtained at positions $(2p-1, 2q)$, $(2p, 2q-1)$ are not destroyed by the transformations involving R^3_{pq} and R^4_{pq}. It follows that

$$S(A')^2 = S(A)^2 - 2\|A_{pq}\|^2.$$

It is in this sense that A moves closer to its Murnaghan form with each Paardekooper

step. For a more detailed description of the Jacobi annihilator $\Im_{pq}$ see [2, 3].

To simplify notation we set (cf. [8, 2]).

$$A_{pp} = \begin{bmatrix} 0 & \alpha \\ -\alpha & 0 \end{bmatrix}, \qquad A_{qq} = \begin{bmatrix} 0 & \beta \\ -\beta & 0 \end{bmatrix},$$

$$A_{pq} = \begin{bmatrix} x & v \\ w & y \end{bmatrix}.$$

Then the angles ϕ_1 and ϕ_2 are computed from the formulas

$$\tan 2\phi_1 = \frac{2(\alpha w - \beta v)}{\alpha^2 - \beta^2 + v^2 - w^2}, \qquad (4)$$

$$\tan \phi_2 = -\frac{v \cos\phi_1 - \beta \sin\phi_1}{\alpha \cos\phi_1 + w \sin\phi_1}, \qquad (5)$$

where $-\frac{\pi}{4} < \phi_1 \leq \frac{\pi}{4}$ and $-\frac{\pi}{2} < \phi_2 \leq \frac{\pi}{2}$. Suppose that A_{pp}, A_{pq} and A_{qq} are changed by the transformations involving R^1_{pq} and R^2_{pq} (half step) into

$$\tilde{A}_{pp} = \begin{bmatrix} 0 & \tilde\alpha \\ -\tilde\alpha & 0 \end{bmatrix}, \tilde{A}_{qq} = \begin{bmatrix} 0 & \tilde\beta \\ -\tilde\beta & 0 \end{bmatrix},$$

$$\tilde{A}_{pq} = \begin{bmatrix} x & 0 \\ 0 & y \end{bmatrix}.$$

Note that v, w have been annihilated and x, y have not been changed. The angles ϕ_3 and ϕ_4 are computed from the formulas

$$\tan 2\phi_3 = \frac{2(\tilde\alpha y + \tilde\beta x)}{\tilde\alpha^2 - \tilde\beta^2 + x^2 - y^2}, \qquad (6)$$

$$\tan \phi_4 = -\frac{x \cos\phi_3 + \tilde\beta \sin\phi_3}{\tilde\alpha \cos\phi_3 + y \sin\phi_3}, \qquad (7)$$

where $-\frac{\pi}{4} < \phi_3 \leq \frac{\pi}{4}$ and $-\frac{\pi}{2} < \phi_2 \leq \frac{\pi}{2}$. For more thorough angle formulas see [8, 2].

Besides the angle determination we have to specify a pivot strategy in which a method of choosing the order of pivot submatrices is given. In this paper we confine ourselves to the row-cyclic strategy. That is, pivot submatrices are chosen in natural order by rows. By "cyclic" we mean to repeat the same procedure after all pivot submatrices are exhausted. One cyclic procedure is called a "cycle". We note that several well known parallel cyclic strategies are closely related to the row-cyclic strategy [6], so phenomena linked with the row-cyclic strategy are relevant for those parallel ones.

The angle formulas together with the row-cyclic strategy define the row-cyclic Paardekooper method for skew-symmetric matrices.

Almost Murnaghan Form

We denote the eigenvalues of A by $\pm i\nu_j$, $\nu_j \geq 0$, $j = 1, \ldots, k$. Let 3δ be the minimum distance among all two distinct eigenvalues of A. We enumerate $\nu_j \geq 0$, $j = 1, \ldots, k$,

$$
\begin{aligned}
\nu_1 &= \quad \ldots \quad = \nu_{s_1}, \\
\nu_{s_1+1} &= \quad \ldots \quad = \nu_{s_2}, \\
&\quad \ldots \\
\nu_{s_{r-1}+1} &= \quad \ldots \quad = \nu_{s_r},
\end{aligned}
$$

and ν_{s_j}, $j = 1, \ldots, r$, are all distinct. If $S(A) < \delta$, we say that A is *almost in Murnaghan form*. In this case we say that the block A_{pp}, $p = 1, \ldots, k$, is affiliated with $\pm i\nu_{s_j}$ (or simply with ν_{s_j}) if $||a_{2p-1,2p}| - \nu_{s_j}| < \delta$. From an eigenvalue perturbation theory (by the Hoffman and Wielandt Theorem [5]) it follows that each block A_{pp}, $p = 1, \ldots, k$, is affiliated with a unique value ν_{s_j} for some $j \in \{1, \ldots, r\}$. Then we have the following result proved in [4]

Theorem 1 *If $\epsilon = S(A) < \delta$ and A_{pp} and A_{qq} $(p < q)$ are affliated with the same $\nu_{s_j} > 0$, we have*

$$\sigma_q a_{2p-1,2q} - \sigma_p a_{2p,2q-1} = O(\epsilon^2), \quad (8)$$

$$a_{2p-1,2q-1} + \sigma_p \sigma_q a_{2p,2q} = O(\epsilon^2), \quad (9)$$

$$|a_{2p-1,2p}| - \nu_{s_j} = O(\epsilon^2), \quad (10)$$

$$|a_{2q-1,2q}| - \nu_{s_j} = O(\epsilon^2), \quad (11)$$

where $\sigma_i = signum(a_{2i-1,2i})$, $1 \leq i \leq k$, and $signum(x) = 1$ if $x \geq 0$ and -1 otherwise. ∎

To be precise in the relations (8)–(11) $O(\epsilon^2)$ actually means $O(\epsilon^m)$, $m \geq 2$. Theorem 1 reveals a special structure of a skew-symmetric matrix almost in Murnaghan form. We note that the relations (8) and (9) do not force $\|A_{pq}\|$ to be $O(\epsilon^2)$. In general we have $\|A_{pq}\| = O(\epsilon)$. Later we will see it through an example.

Application to the Paardekooper Method

It is known (see [3]) that a skew-symmetric matrix almost in Murnaghan form converges quadratically when we apply the row-cyclic Paardekooper method, provided that the eigenvalues of A are at most double and the diagonal blocks which are affiliated with the same value are located in successive positions. By quadratic convergence we mean $S(\bar{A}) \leq cS(A)^2$ for some constant c, where $\bar{A}$ stands for the matrix obtained from A after a cycle. But the method fails to converge quadratically when multiplicities of some nonzero eigenvalues of A are larger than two. To analyse such a behaviour we need the following lemma proved in [4] using the relations (8)–(11) together with the angle formulars (4)–(7).

Lemma 2 *If A_{pp} and A_{qq} $(p < q)$ are affiliated with the same $\nu_{s_j} > 0$, we have*

$$\phi_1, \ \phi_2, \ \phi_3, \ \phi_4 = O(1) \ as \ S(A) \to 0.$$

where ϕ_1, ϕ_2, ϕ_3 and ϕ_4 are angles used in the Jacobi annihilator $\mathfrak{S}_{pq}$. ∎

Now we can argue the failure of the quadratic convergence of the row-cyclic Paardekooper method under the presence of nonzero eigenvalues who multiplicities are greater than two. Let

$$
A = \begin{bmatrix}
A_{11} & A_{12} & A_{13} & A_{14} \\
A_{21} & A_{22} & A_{23} & A_{24} \\
A_{31} & A_{32} & A_{33} & A_{34} \\
A_{41} & A_{42} & A_{43} & A_{44}
\end{bmatrix}
$$

be a given 8 x 8 skew-symmetric matrix almost in Murnaghan form. We further assume that A_{11}, A_{22} and A_{33} are affiliated with the same $\nu_{s_1} > 0$, and A_{44} is affiliated with ν_{s_2}. Suppose we apply the row-cyclic Paardekooper method to A. Because of the row-cyclic strategy the first pivot submatrix is A_{12}. After the annihilation of A_{12} we have $A'_{12} = O$, that is, we have

$$
A' = \begin{bmatrix}
A'_{11} & O & A'_{13} & A'_{14} \\
O & A'_{22} & A'_{23} & A'_{24} \\
A'_{31} & A'_{32} & A_{33} & A_{34} \\
A'_{41} & A'_{42} & A_{43} & A_{44}
\end{bmatrix}
$$

Note that A_{33}, A_{34}, A_{43} and A_{44} have not changed. Because of the row-cyclic strategy A'_{13} is the next pivot submatrix. When A'_{13} is annihilated, the zero block in A' gets contributions from the elements of A'_{23} by the skew-symmetry. Since in general $\|A'_{23}\| = O(\epsilon)$, where $\epsilon = S(A)$, and the angles involved in $\mathfrak{S}_{13}$ can be $O(1)$ by Lemma 2, the size of the contribution to the zero block can be $O(\epsilon) \cdot O(1) = O(\epsilon)$. This means that there might be a 2 x 2 submatrix which becomes to the order of $S(A)$ after its annihilation. This implies the failure of the quadratic convergence.

What we need to do is to construct an example which demonstrates our speculation. We now demonstrate how Maple can be used for this purpose.

Constructing and Analysing an Example using Maple

To generate skew-symmetric matrix A of order 8 we start from the following Murnaghan form M:

$$M = \begin{bmatrix} 0 & 1 & & & & & & \\ -1 & 0 & & & & & & \\ & & 0 & 1 & & & & \\ & & -1 & 0 & & & & \\ & & & & 0 & 1 & & \\ & & & & -1 & 0 & & \\ & & & & & & 0 & 2 \\ & & & & & & -2 & 0 \end{bmatrix},$$

where all unspecified elements are zero. Note that $\pm i$ are the nonzero eigenvalues of M of multiplicity 3, and $\pm 2i$ are simple eigenvalues. We have generated A by applying the following orthogonal similarity transformation to M:

$$A = Q^T M Q$$

where $Q = P\, R(5,7;\phi)\, R(2,8;\phi)$ with $\phi = \arcsin t$ (recall (3)) and

$P =$

$$\frac{1}{\sqrt{1+t^2}} \begin{bmatrix} 0 & 1 & 0 & 0 & 0 & 0 & 0 & -t \\ 1 & 0 & 0 & 0 & 0 & -t & 0 & 0 \\ 0 & 0 & 0 & 1 & -t & 0 & 0 & 0 \\ 0 & 0 & 1 & 0 & 0 & 0 & -t & 0 \\ 0 & 0 & 0 & t & 1 & 0 & 0 & 0 \\ t & 0 & 0 & 0 & 0 & 1 & 0 & 0 \\ 0 & 0 & t & 0 & 0 & 0 & 1 & 0 \\ 0 & t & 0 & 0 & 0 & 0 & 0 & 1 \end{bmatrix}$$

Note that P is orthogonal, and hence A is orthogonally similar to M. This means that A is skew-symmetric and that $\pm i$ are the nonzero eigenvalues of A of multiplicity 3 and $\pm 2i$ are simple eigenvalues of A.

Note that

$R(5,7;\phi) =$

$$\begin{bmatrix} 1 & 0 & 0 & 0 & 0 & 0 & 0 & 0 \\ 0 & 1 & 0 & 0 & 0 & 0 & 0 & 0 \\ 0 & 0 & 1 & 0 & 0 & 0 & 0 & 0 \\ 0 & 0 & 0 & 1 & 0 & 0 & 0 & 0 \\ 0 & 0 & 0 & 0 & \sqrt{1-t^2} & 0 & t & 0 \\ 0 & 0 & 0 & 0 & 0 & 1 & 0 & 0 \\ 0 & 0 & 0 & 0 & -t & 0 & \sqrt{1-t^2} & 0 \\ 0 & 0 & 0 & 0 & 0 & 0 & 0 & 1 \end{bmatrix}$$

and

$R(2,8;\phi) =$

$$\begin{bmatrix} 1 & 0 & 0 & 0 & 0 & 0 & 0 & 0 \\ 0 & \sqrt{1-t^2} & 0 & 0 & 0 & 0 & 0 & t \\ 0 & 0 & 1 & 0 & 0 & 0 & 0 & 0 \\ 0 & 0 & 0 & 1 & 0 & 0 & 0 & 0 \\ 0 & 0 & 0 & 0 & 1 & 0 & 0 & 0 \\ 0 & 0 & 0 & 0 & 0 & 1 & 0 & 0 \\ 0 & 0 & 0 & 0 & 0 & 0 & 1 & 0 \\ 0 & -t & 0 & 0 & 0 & 0 & 0 & \sqrt{1-t^2} \end{bmatrix}$$

We can declare M, P, $R(5,7;\phi)$ and $R(2,8;\phi)$ by using Maple. We provide Maple code for M and $R(5,7;\phi)$, since others can be declared similarly. In the code $R57$ stands for $R(5,7;\phi)$.

```
> with(linalg):
> n := 8 :
> # We declare M.
> M := matrix(n, n, (i, j)- > 0) :
> M[1, 2] := 1 :
> M[3, 4] := 1 :
```

```
> M[5,6] := 1 :
> M[7,8] := 2 :
> M[2,1] := -1 :
> M[4,3] := -1 :
> M[6,5] := -1 :
> M[8,7] := -2 :
> # We declare R(5,7;φ).
> R57 := matrix(n,n,(i,j)- > 0) :
> for i to n do R57[i,i] := 1 od:
> R57[5,5] := sqrt(1 - t^2) :
> R57[5,7] := t :
> R57[7,7] := sqrt(1 - t^2) :
> R57[7,5] := -t :
```

Then we can generate A by the following segment of Maple code:

```
> A := multiply(M, P) :
> A := multiply(transpose(P), A) :
> A := multiply(A, R57) :
> A := multiply(transpose(R57), A) :
> A := multiply(A, R28) :
> A := multiply(transpose(R28), A) :
```

Using the partition $A = \begin{bmatrix} \mathcal{A}_{11} & \mathcal{A}_{12} \\ -\mathcal{A}_{12}^T & \mathcal{A}_{22} \end{bmatrix}$, where $\mathcal{A}_{11}$ is 6 x 6, Maple returns the following elements of A:

Note that $\delta = 1/3$ and $S(A) = O(t)$ as $t \to 0$. Hence the condition $S(A) < \delta$ is fulfilled for sufficiently small t. Under that condition we see that A_{11}, A_{22}, A_{33} and A_{44} are affiliated with 1 and 2, respectively.

The structure of $\mathcal{A}_{11}$ demonstrates the validity of Theorem 1. By the following segment of Maple code we verify the validity of the assertions in the relations (8) and (9) of Theorem 1. In the code $sg1$, $sg2$ and $sg3$ stand for $signum(a_{12})$, $signum(a_{34})$ and $signum(a_{56})$, respectively. Obviously $sg1 = sg2 = -1$ and $sg3 = 1$ for small t.

```
> sg1 := -1 :
> sg2 := -1 :
> sg3 := 1 :
> b12 := sg2 * A[1,4] - sg1 * A[2,3] :
> b12 := series(b12, t = 0, 2);
        b12 := O(t²)
> c12 := A[1,3] + sg1 * sg2 * A[2,4] :
> c12 := series(c12, t = 0, 2);
        c12 := 0
> b13 := sg3 * A[1,6] - sg1 * A[2,5] :
> b13 := series(b13, t = 0, 2);
        b13 := O(t⁴)
> c13 := A[1,5] + sg1 * sg3 * A[2,6] :
> c13 := series(c13, t = 0, 2);
        c13 := O(t³)
```

$$\mathcal{A}_{11} = \begin{bmatrix} 0 & -\frac{\sqrt{1-t^2}+t^2}{1+t^2} & 0 & -\frac{t^2}{1+t^2} & -\frac{t\sqrt{1-t^2}}{1+t^2} & 0 \\ & 0 & \frac{2t^2(1-\sqrt{1-t^2})}{1+t^2} & 0 & \frac{2t^2(\sqrt{1-t^2}-1)}{1+t^2} & -\frac{t(\sqrt{1-t^2}+t^2)}{1+t^2} \\ & & 0 & -\frac{1}{1+t^2} & \frac{t\sqrt{1-t^2}}{1+t^2} & 0 \\ & & & 0 & \frac{t^2}{1+t^2} & \frac{t}{1+t^2} \\ & \text{skew-sym.} & & & 0 & \frac{\sqrt{1-t^2}}{1+t^2} \\ & & & & & 0 \end{bmatrix},$$

$$\mathcal{A}_{12} = \begin{bmatrix} -\frac{t^2}{1+t^2} & \frac{t(\sqrt{1-t^2}-1)}{1+t^2} \\ \frac{2t\sqrt{1-t^2}(1-\sqrt{1-t^2})}{1+t^2} & 0 \\ \frac{t^2}{1+t^2} & \frac{2t(\sqrt{1-t^2}+t^2)}{1+t^2} \\ -\frac{t\sqrt{1-t^2}}{1+t^2} & 0 \\ \frac{t^2}{1+t^2} & -\frac{2t(\sqrt{1-t^2}+t^2)}{1+t^2} \\ -\frac{t}{1+t^2} & \frac{t^2(1-\sqrt{1-t^2})}{1+t^2} \end{bmatrix}, \quad \mathcal{A}_{22} = \begin{bmatrix} 0 & \frac{2\sqrt{1-t^2}(\sqrt{1-t^2}+t^2)}{1+t^2} \\ \text{skew-s.} & 0 \end{bmatrix}.$$

```
> b23 := sg3 * A[3, 6] − sg2 * A[4, 5] :
> b23 := series(b23, t = 0, 2);
         b23 := O(t²)
> c23 := A[3, 5] + sg2 * sg3 * A[4, 6] :
> c23 := series(c23, t = 0, 2);
         c23 := O(t³)
```

Note that the output of Maple demonstrates the validity of the assertions in the relations (8) and (9) of Theorem 1. It is summarized in the following Table:

(p, q)	$\sigma_q a_{2p-1,2q} - \sigma_p a_{2p,2q-1}$	$a_{2p-1,2q-1} + \sigma_p \sigma_q a_{2p,2q}$
$(1, 2)$	$O(t^2)$	0
$(1, 3)$	$O(t^4)$	$O(t^3)$
$(2, 3)$	$O(t^2)$	$O(t^3)$

The following output of Maple code also demonstrates the validity of the assertions in the relations (10) and (11) of Theorem 1.

```
> b11 := abs(A[1, 2]) − 1 :
> b11 := series(b11, t = 0, 2);
         b11 := O(t²)
> b22 := abs(A[3, 4]) − 1 :
> b22 := series(b22, t = 0, 2);
         b22 := O(t²)
> b33 := abs(A[5, 6]) − 1 :
> b33 := series(b33, t = 0, 2);
         b33 := O(t²)
```

We also note that $\|A_{13}\|$, $\|A_{23}\| = O(t)$, which was asserted right after Theorem 1. This is verified by the following segment of Maple code:

```
> size13 := sqrt(A[1, 5]^2 + A[1, 6]^2
>>     +A[2, 5]^2 + A[2, 6]^2) :
> size13 := series(size13, t = 0, 2);
         size13 := O(t)
> size23 := sqrt(A[3, 5]^2 + A[3, 6]^2
>>     +A[4, 5]^2 + A[4, 6]^2) :
> size23 := series(size23, t = 0, 2);
         size23 := O(t)
```

Now suppose we apply the row-cyclic Paardekooper method to A. By $A^{(r)} = \left(a_{ij}^{(r)} \right)$ we denote the matrix obtained from A after r rotational transformations and by ϕ_{ij} the angle used in the rotation matrix in the (i, j) plane. We also denote $\tan \phi_{ij}$, $\sin \phi_{ij}$ and $\cos \phi_{ij}$ by t_{ij}, s_{ij} and c_{ij}, respectively.

Because of the row-cyclic strategy the first pivot submatrix is A_{12}. From the formula (4) we obtain

$$\tan(2\phi_{13}) = \tag{12}$$
$$2 \frac{a_{12}a_{23} - a_{34}a_{14}}{(a_{12})^2 - (a_{34})^2 + (a_{14})^2 - (a_{23})^2}.$$

The following segment of Maple code estimates ϕ_{13}, c_{13} and s_{13} using the relation (12). In the code $phi13$, $c13$, $s13$ stand for ϕ_{13}, c_{13}, s_{13}, respectively.

```
> ex1 := 2 * (A[1, 2] * A[2, 3] − A[3, 4] * A[1, 4])/
>> (A[1, 2]^2 − A[3, 4]^2
>> +A[1, 4]^2 − A[2, 3]^2) :
> ex1 := series(ex1, t = 0) :
> ex1 := simplify(ex1) :
> phi13 := arctan(ex1)/2 :
> phi13 := series(phi13, t = 0) :
> phi13 := simplify(phi13) :
> evalf(phi13);
          −0.5535743590 + O(t^4)
> c13 := cos(phi13) :
> c13 := series(c13, t = 0) :
> c13 := simplify(c13) :
> s13 := sin(phi13) :
> s13 := series(s13, t = 0) :
> s13 := simplify(s13) :
```

Note that the output of ϕ_{13} demonstrates the validity of Lemma 2.

Using the formular (5) we have

$$t_{24} = -\frac{a_{14}c_{13} - a_{34}s_{13}}{a_{12}c_{13} + a_{23}s_{13}}. \quad (13)$$

The following segment of Maple code estimates c_{24} and s_{24} using the relation (13).

```
> t24 := −(A[1, 4] * c13 − A[3, 4] * s13)/
>> (A[1, 2] * c13 + A[2, 3] * s13) :
> t24 := simplify(series(t24, t = 0)) :
> phi24 := simplify(series(arctan(t24), t = 0)) :
> c24 := simplify(series(cos(phi24), t = 0)) :
> s24 := simplify(series(sin(phi24), t = 0)) :
```

Note that $\phi_{14} = \phi_{23} = 0$, since $a_{13}^{(2)} = a_{13} = 0$ and $a_{24}^{(2)} = a_{24} = 0$. This means that we do not need to consider the transformations involving $R(1, 4; \phi_{14})$ and $R(2, 3; \phi_{23})$, because they are identity transformation.

Because of the row-cyclic strategy $A_{13}^{(4)}$ is the next pivot submatrix. Using the formula (4) we have

$$\tan(2\phi_{15}) = \quad (14)$$
$$2 \frac{a_{12}^{(4)}a_{25}^{(4)} - a_{56}^{(4)}a_{16}^{(4)}}{(a_{12}^{(4)})^2 - (a_{56}^{(4)})^2 + (a_{16}^{(4)})^2 - (a_{25}^{(4)})^2},$$

where one can verify

$$a_{12}^{(4)} = (a_{12}c_{13} + a_{23}s_{13})c_{24} \\ - (a_{14}c_{13} - a_{34}s_{13})s_{24}. \quad (15)$$

$$a_{25}^{(4)} = a_{25}c_{24} - a_{45}s_{24}. \quad (16)$$

$$a_{16}^{(4)} = a_{16}, \quad a_{56}^{(4)} = a_{56}. \quad (17)$$

The following segment of Maple code estimates s_{15} using the relations (14)–(17). In the code aij_4 denote $a_{ij}^{(4)}$:

```
> a12_4 := (A[1, 2] * c13 + A[2, 3] * s13) * c24
>> −(A[1, 4] * c13 − A[3, 4] * s13) * s24 :
> a12_4 := simplify(series(a12_4, t = 0)) :
> a25_4 := A[2, 5] * c24 − A[4, 5] * s24 :
> a25_4 := simplify(series(a25_4, t = 0)) :
> a16_4 := simplify(series(A[1, 6], t = 0)) :
> a56_4 := simplify(series(A[5, 6], t = 0)) :
> ex2 := 2 * (a12_4 * a25_4 − a56_4 * a16_4)/
>> (a12_4^2 − a56_4^2 + a16_4^2 − a25_4^2) :
> ex2 := simplify(series(ex2, t = 0)) :
> phi15 := arctan(ex2)/2 :
> phi15 := simplify(series(phi15, t = 0)) :
> s15 := simplify(series(sin(phi15), t = 0)) :
```

During the transformation involving $R(1, 5; \phi_{15})$, $(1, 2)$ block gets contributions from $(2, 3)$ block by skew-symmetry. In particular

$$a_{13}^{(5)} = a_{13}^{(4)}c_{15} + a_{35}^{(4)}s_{15} = a_{35}^{(4)}s_{15}, \quad (18)$$

since $a_{13}^{(4)} = 0$, where one can verify that

$$a_{35}^{(4)} = a_{35}c_{13} + a_{15}s_{13}. \quad (19)$$

The final segment of Maple code estimates $a_{13}^{(5)}$ using the relations (18) and (19):

```
> a35_4 := A[3, 5] * c13 + A[1, 5] * s13 :
> a35_4 := simplify(series(a35_4, t = 0)) :
> a13_5 := a35_4 * s15 :
> a13_5 := simplify(series(a13_5, t = 0)) :
> a13_5 := evalf(a13_5);
      a13_5 := −0.2612289826 t + O(t^3)
```

From the output we see that $a_{13}^{(5)} = O(t)$. Since $S(A) = O(t)$, we conclude that $(1, 2)$ block becomes the order of $S(A)$ by later rotational transformation after its annihilation. So this example demonstrates the failure of the quadratic convergence of the row-cyclic Paardekooper method in the presence of nonzero eigenvalues whose multiplicities are greater than two.

We note that in constructing and analysing the example without Maple the calculations would have been extremely tedious. For example, from the relation (12)

$$\tan(2\phi_{13}) = \frac{-2[2t^2(t^2 + \sqrt{1-t^2})(1 - \sqrt{1-t^2}) + t^2]}{(t^2 + \sqrt{1-t^2})^2 - 1 + t^4 - 4t^4(1 - \sqrt{1-t^2})^2}.$$

Clearly it is a tedious job to find the Taylor series of the above expression with $O(t^6)$ of truncation by hands. To analyse our example we need to do such computations over and over. It is precisely in this point where Maple comes into play to help us.

References

[1] K. K. Gupta: On a Combined Strum Sequence and Inverse Iteration Technique for Eigenproblem of Spinning Structures, International Journal for Numerical Methods in Engineering 7: 509–518 (1973).

[2] V. Hari: On Quadratic Convergence of the Paardekooper Method I, Glasnik Matematički 67: 183–195, 1982.

[3] V. Hari and N. H. Rhee: On Quadratic Convergence of the Paardekooper Method II, accepted for publication in Glasnik Matematički.

[4] V. Hari and N. H. Rhee: A Matrix Pair of an Almost Diagonal Skew-Symmetric and Symmetric Positive Definite Matrix, accepted for publication in Linear Algebra and Its Applications.

[5] A. J. Hoffman and H. W. Wielandt: The Variation of the Spectrum of a Normal Matrix, Duke Math. J. 20: 37–39, 1953.

[6] F. Luk and H. Park: On the Equivalence and Convergence of Parallel Jacobi SVD Algorithms, Proc. SPIE, Vol. 826, Advanced Algorithms and Architectures for Signal Processing: 152–159, 1987.

[7] F. D. Murnaghan and A. Winter: A Canonical Form for Real Matrices under Orthogonal Transformations, USA Nat. Acad. Sci. 17: 417–420, 1931.

[8] M. H. C. Paardekooper: An Eigenvalue Algorithm for Skew-symmetric Matrices, Numer.Math. 17: 189–202, 1971.

[9] R. A. Rosanoff et al. : Numerical Conditions of Stiffness Matrix Formulations for Frame Structures, Proceedings of the 2nd Conference on Matrix Methods in Structural Mechanics, WPAFB Dayton, Ohio, 1968.

The author Noah H. Rhee studied Mathematics at Seoul National University and got his Ph.D. in 1987 from Michigan State University. He is presently an Assistant Professor at the University of Missouri–Kansas City, and his current interests include numerical linear algebra and numerical analysis. He can be reached by the following address:

Noah H. Rhee
Department of Mathematics
University of Missouri–Kansas City
Kansas City, MO 64110-2499

E-Mail: rhee@vax1.umkc.edu
Tel: (816) 235-2854

AN ALGORITHM TO COMPUTE FLOATING POINT GROEBNER BASES

Kiyoshi Shirayanagi
NTT Communication Science Laboratories, Kyoto, Japan

1 Introduction

Gröbner basis (GB) techniques are a valuable tool for solving many problems in polynomial ideal theory. As is well known, the process of computing a GB may involve large numbers of intermediate coefficients - say from a field k - even when the final GB does not involve many coefficients. In fact, the cost of performing exact arithmetic in k with the intermediate coefficients is a major factor determining the computational cost of computing the GB. This paper proposes a new approach using floating point computation that can be applied when k is a subfield of the real numbers.[1]

Basically we mimic Buchberger's algorithm in [3]. However, the big question then would be "how small must floating point coefficients be, to be considered zero?". The subject of this paper is to propose a criterion for answering this question. Our key idea is to calculate and keep track of an error term for every coefficient that occurs at each step of the S-polynomial calculation and polynomial reduction, and to judge coefficients as zero by estimation of their accumulated errors. To keep track of the errors, bracket coefficients for polynomials are introduced. These are based on floating point numbers together with error terms. The arithmetic for bracket coefficients is a kind of *interval arithmetic* (see, for example, [8], [1], and [9]).

The proposed algorithm gives a sort of *approximate* GB (or AGB), not a *true* GB. There are a number of problems which are typically solved by computing a GB for which it is only necessary to compute an AGB. Moreover, for numerous examples, it is much faster to compute an AGB than a GB.

In Section 2, a new notion of convergence of a sequence of floating point polynomials and an approximate GB are defined to describe what our algorithm does. We present an example illustrating why the most naive approach does not work. In addition, possible applications are mentioned. In Section 3, basic notions for our algorithm are provided as well as a key theorem. Bracket coefficient polynomials are introduced. They play a central role in this paper. The key theorem presents a criterion for deciding whether a true coefficient is zero, in terms of bracket coefficients. In Section 4, the algorithm is described and the termination and correctness of it are proved. In Section 5, examples of running the algorithm are presented. The algorithm is implemented in Maple V/Sun SPARC. In Section 6, the experimental results are discussed by comparing them to those of the conventional algorithm. Finally, an important open problem is given.

Throughout this paper, we assume that the set of all floating point numbers is contained in the real field **R**. When we simply say a polynomial, it denotes a floating point polynomial or a real polynomial in variables $x_1, \ldots, x_n$.

---

[1]While our approach applies to the case that k is the field of rational numbers, several other approaches apply there as well. We refer the reader to p-adic or modular approaches as found in [11], [12], [10].

2 Floating Point Gröbner Sequence

In this section, we introduce a sequence of sets of floating point polynomials that converges to a true Gröbner basis in a strong way. First of all, we define the monomial support of a polynomial or a finite set of polynomials.

Mathematical Computation with Maple V:
Ideas and Applications
Tom Lee, Editor
©1993 Birkhäuser Boston

Definition 1 (Monomial Support) *The monomial support of a polynomial*

$$f = \sum_{i_1,\ldots,i_n} a_{i_1,\ldots,i_n} x_1^{i_1} \cdots x_n^{i_n}$$

is the set of power products

$$\{x_1^{i_1} \cdots x_n^{i_n} \mid a_{i_1,\ldots,i_n} \neq 0\},$$

denoted $\mathcal{M}(f)$. *The monomial support of a finite set* $F = \{f_1, \ldots, f_n\}$ *of polynomials is the set of subsets of the set of power products:* $\{\mathcal{M}(f_1), \ldots, \mathcal{M}(f_n)\}$, *denoted* $\mathcal{M}(F)$.

Note that $\mathcal{M}(0) = \emptyset$.

Example. For $f = 2.99x^2 + .999xy^2 - 1.002z + .249$, $\mathcal{M}(f) = \{x^2, xy^2, z, 1\}$. For $F = \{3x^2 + xy^2 - z + 1/4, 1/3x + y^2 + 1/2, x^2z - 1/2x - y^2\}$, $\mathcal{M}(F) = \{\{x^2, xy^2, z, 1\}, \{x, y^2, 1\}, \{x^2z, x, y^2\}\}$.

Next we discuss the general notion of (coefficientwise) convergence and the specialized notion of supportwise convergence which we require.

The natural definition of the coefficientwise convergence of a sequence of polynomials is given by the following.

Definition 2 (Coefficientwise Convergence) *Let* $\{f_\nu\}_\nu$ *be a sequence of polynomials and* f *be a polynomial. Then,* f_ν *coefficientwise converges to* f *($f_\nu \to f$ coefficientwise) iff* $f_\nu = \sum_{i_1,\ldots,i_n} a_{i_1,\ldots,i_n}^\nu x_1^{i_1} \cdots x_n^{i_n}$ *and* $f = \sum_{i_1,\ldots,i_n} a_{i_1,\ldots,i_n} x_1^{i_1} \cdots x_n^{i_n}$ *such that* $\lim_{\nu\to\infty} a_{i_1,\ldots,i_n}^\nu = a_{i_1,\ldots,i_n}$ *for all* $i_1, \ldots, i_n$.

The stronger property of supportwise convergence is one of the central themes of this paper:

Definition 3 (Supportwise Convergence) *Let* $\{f_\nu\}_\nu$ *be a sequence of polynomials and* f *be a polynomial. Then,* f_ν *supportwise converges to* f *($f_\nu \to f$ supportwise) iff there is an* N *such that* (1) $\mathcal{M}(f_\nu) = \mathcal{M}(f)$ *for all* $\nu \geq N$, *and* (2) $\{f_\nu\}_{\nu \geq N}$ *is coefficientwise convergent to* f.

Moreover let $\{F_\nu\}_\nu$ *be a sequence of finite sets of polynomials and* F *be a finite set of polynomials. Then,* F_ν *supportwise converges to* F *($F_\nu \to F$ supportwise) iff* $F_\nu = \{f_1^\nu, \ldots, f_n^\nu\}$ *and* $F = \{f_1, \ldots, f_n\}$ *such that* $f_i^\nu \to f_i$ *supportwise for all* $i \in [1, n]$.

Example. Let $f_1^\nu = \frac{\nu}{\nu+1}x^3y - \frac{\nu-1}{\nu}x + 1$ and $f_2^\nu = \frac{\nu-1}{\nu}x^2y^2 - y + \frac{2\nu}{\nu+1}$. Then $f_1^\nu \to f_1 = x^3y - x + 1$ and

$f_2^\nu \to f_2 = x^2y^2 - y + 2$ supportwise. Also $F_\nu = \{f_1^\nu, f_2^\nu\} \to F = \{f_1, f_2\}$ supportwise. However, for $f_3^\nu = \frac{\nu}{\nu+1}x^3y - \frac{1}{\nu}x + 1$ and $f_3 = x^3y + 1$, $f_3^\nu \to f_3$ coefficientwise but $f_3^\nu \not\to f_3$ supportwise, since $\mathcal{M}(f_3^\nu) \neq \mathcal{M}(f_3)$ for any ν.

The aim of this paper is to provide an algorithm that computes a set G_μ for each μ where $\{G_\mu\}_\mu$ is supportwise convergent to a true Gröbner basis G. Let us call such a sequence $\{G_\mu\}_\mu$ a *floating point Gröbner sequence*. If $\{G_\mu\}_\mu$ is any floating point Gröbner sequence, there is an M where $\mathcal{M}(G_\mu) = \mathcal{M}(G)$ for all $\mu \geq M$. For $\mu \geq M$ we shall call G_μ a *floating point Gröbner basis* with precision μ, or simply an *approximate Gröbner basis*.

Of course there is an obvious way to construct a floating point Gröbner sequence. That is, (1) Compute a true Gröbner basis G by the conventional Buchberger's algorithm ([3]), and then (2) For each μ, truncate the coefficients of G to precision μ. However, our algorithm avoids exact arithmetic and the associated memory requirements. On the other hand, we must be able to compute with arbitrarily high precision coefficients.

For this purpose one naive idea may be first to take F_μ that is the floating point approximation of F in coefficients to precision μ, and then to apply Buchberger's algorithm to F_μ. Such an approach fails before the GB computation begins. This is seen by the trivial example $\{3x - 1, x - 1/3\}$. With any truncation of $1/3$, the first polynomial is no longer a scalar multiple of the second one, and the ideal generated by them has changed to $\mathbf{R}[x]$. See Section 5 for the result of our algorithm for a similar and less trivial example.

Let V be the quotient algebra

$$\mathbf{R}[x_1, \ldots, x_n]/Ideal(F),$$

where $Ideal(F)$ is the ideal generated by F. Once an approximate Gröbner basis G_μ is obtained, the following problems can be solved *not approximately but strictly* by the conventional methods, since $\mathcal{M}(G_\mu) = \mathcal{M}(G)$ for a true Gröbner basis G. (1) Decide whether V is finite-dimensional or not as $\mathbf{R}$ vector space, (2) Compute an $\mathbf{R}$-basis of V, (3) Decide whether the system F of polynomials is solvable or not, etc.

Moreover, the system F may be *approximately* solved, since every coefficient of G_μ converges to the corresponding one of G. See [3] (Methods 6.6,

96

6.8, 6.9, and 6.10 etc.) for the details on the conventional methods.

3 Theoretical Foundation

3.1 Basic Notions

It is natural to mimic Buchberger's algorithm but with floating point coefficients. However, the largest problem is *"how small must coefficients be for us to consider them to be zero?"*. It should be noted that *if our algorithm can judge a coefficient as zero iff the corresponding coefficient in Buchberger's algorithm is truly zero, then the results are exactly the same in monomial support.* Thus we must provide a useful criterion for this zero judgement. Our idea is to calculate and keep track of an error of every coefficient that occurs at each step of the S-polynomial calculation and polynomial reduction.

To keep track of the errors, a *bracket coefficient polynomial*, or simply a *BC polynomial* with a given precision μ is introduced. This is 0 or a polynomial of the form

$$\sum_{i_1,\ldots,i_n} [A_{i_1,\ldots,i_n}, \alpha_{i_1,\ldots,i_n}] x_1^{i_1} \cdots x_n^{i_n},$$

where $A_{i_1,\ldots,i_n}$ and $\alpha_{i_1,\ldots,i_n}$ are floating point numbers with precision μ. $A_{i_1,\ldots,i_n}$ and $\alpha_{i_1,\ldots,i_n}$ are intended to be an approximation of a true coefficient and its error, respectively. A special symbol **0** is permitted in place of $[A_{i_1,\ldots,i_n}, \alpha_{i_1,\ldots,i_n}]$.

Next we have to define S-BC polynomial and BC-reduction for BC polynomials. In this paper, using 10 as the base, a floating point number A with precision μ is expressed by

$$A = \pm .a_1 a_2 \ldots a_\mu \times 10^{e(A)},$$

where $1 \le a_1 \le 9$, $0 \le a_i \le 9$ $(2 \le i \le \mu)$, and $e(A)$ is the exponent of A. For floating point arithmetic, we apply the *roundoff* manipulation, which means counting .5 and higher fractions as a unit and cutting away the rest. In this paper we assume that transgressions of the range of numbers (underflow and overflow) do not happen. Following [7], we use a notation $\oplus_\mu$, $\ominus_\mu$, and $\otimes_\mu$ (or simply $\oplus$, $\ominus$, and $\otimes$ when there is no confusion) for floating point addition, subtraction, and multiplication with precision μ, respectively. Throughout this paper, floating point division is not considered because of its complicated error analysis.

Furthermore we apply *round up* with precision μ, denoted $\uparrow_\mu$ or simply $\uparrow$, see [1], for the addition or multiplication of errors. That is, $\uparrow (\alpha + \beta)$ (resp. $\uparrow (\alpha\beta)$) is the same as $\alpha \oplus \beta$ (resp. $\alpha \otimes \beta$) except that it rounds the $(\mu + 1)$-th fraction upward-directedly to 10.

Note that floating point (and round up) addition, subtraction, and multiplication with fixed precision have difficulties that the associative law or distributive law does not necessarily hold in general. Fortunately these deficiencies will not cause a problem for our algorithm.

We define floating point arithmetic for bracket coefficients in BC polynomials. $\uparrow (\alpha + \beta + \gamma + \cdots + \xi)$ denotes $\uparrow (\uparrow (\cdots \uparrow (\uparrow (\alpha + \beta) + \gamma) + \cdots) + \xi)$.

Definition 4 (BC Arithmetic) *With precision μ,*
Addition: $[A, \alpha] \oplus [B, \beta] = [A \oplus B, \uparrow (\alpha + \beta + 5 \times 10^{e(A \oplus B) - (\mu+1)})]$
Subtraction: $[A, \alpha] \ominus [B, \beta] = [A \ominus B, \uparrow (\alpha + \beta + 5 \times 10^{e(A \ominus B) - (\mu+1)})]$
Multiplication: $[A, \alpha] \otimes [B, \beta] = [A \otimes B, \uparrow (\uparrow (\alpha\beta) + \uparrow (\alpha|B|) + \uparrow (\beta|A|) + 5 \times 10^{e(A \otimes B) - (\mu+1)})]$

Remark. In the above addition, if $A \oplus B = 0$, then the term $5 \times 10^{e(A \oplus B) - (\mu+1)}$ in the error is dropped. In the case of general base b of floating point numbers, $5 \times 10^{e(A \oplus B) - (\mu+1)}$ is replaced by $\lceil b/2 \rceil \times b^{e(A \oplus B) - (\mu+1)}$, where $\lceil b/2 \rceil$ is the minimal integer not less than $b/2$. Similar for the subtraction and multiplication.

The adequacy of this definition will be shown to be clear in Lemma 1 later. BC arithmetic may be a kind of *interval arithmetic* ([8]), more precisely *machine* (or *rounded*) *interval arithmetic* ([1]) or *circular arithmetic* ([1], [9]), although the definitions of these arithmetic are slightly different from ours. However, the emphasis of this paper is not on the choice of arithmetic, but on the idea of keeping track of errors and more importantly on the zero judgement of coefficients by estimation of the accumulated errors.

Consider BC polynomials with a given precision μ. We can add, subtract, and multiply such polynomials as well as usual polynomials as follows:

$$[A, \alpha]t + [B, \beta]t = ([A, \alpha] \oplus_\mu [B, \beta])t$$

$$(\text{or } [A, \alpha]t - [B, \beta]t = ([A, \alpha] \ominus_\mu [B, \beta])t),$$

$$\sum_i [A_i, \alpha_i]t_i \cdot \sum_j [B_j, \beta_j]u_j$$

$$= \sum_{i,j} ([A_i, \alpha_i] \otimes_\mu [B_j, \beta_j]) t_i u_j,$$

where t, t_i, u_j are power products. Here for 0, we have

$$f + 0 = 0 + f = f,$$

for any BC polynomial f.

Moreover, as a convention, we apply the following laws:

$$[A, \alpha] \ominus [A, \alpha] = 0,$$

$$0 \cdot t = 0,$$

$$-([A, \alpha]t) = [-A, \alpha]t.$$

Note that in general the set of BC polynomials is not a ring because, as mentioned above, the arithmetic operators are not associative and distributive in the bracket coefficients.

Now we are prepared to define S-BC polynomial and BC-reduction. Given an admissible term ordering, $LP(f)$ denotes the leading power product of a polynomial or BC polynomial f. If we write $f = f' + rest(f)$, then f' is a term of f and $rest(f)$ is the other terms of f.

Definition 5 (S-BC polynomial) *Let f and g be BC polynomials with precision μ $[A, \alpha]LP(f) + rest(f)$ and $[B, \beta]LP(g) + rest(g)$ respectively. And let LCM be $lcm(LP(f), LP(g))$. Then the S-BC polynomial of f and g with precision μ is the BC polynomial*

$$[B, \beta] \cdot \frac{LCM}{LP(f)} \cdot f - [A, \alpha] \cdot \frac{LCM}{LP(g)} \cdot g$$

(with precision μ), denoted $S\text{-}BCpoly_\mu(f, g)$.

Definition 6 (BC-reduction) *Let f be a BC polynomial and F a finite set of BC polynomials with precision μ. Then,*

$$f \xrightarrow{BC}_F h \quad (\text{``} f \text{ BC-reduces to } h \text{ modulo } F\text{''})$$

with precision μ

iff $f \xrightarrow{BC}_{g,u}$ and $h = [B, \beta] \cdot f - [A, \alpha]u \cdot g$

(with precision μ), where $f \xrightarrow{BC}_{g,u}$ ("f is BC-reducible using g and u") iff there are $g \in F$ and a power product u such that $f = [A, \alpha] \cdot u \cdot LP(g) + rest(f)([A, \alpha] \neq 0)$, $g = [B, \beta]LP(g) + rest(g)$.

BC-normal form of f modulo F with precision μ, denoted $BC\text{-}NForm_\mu(f, F)$, is defined using BC-reduction, in a manner similar to the conventional normal form.

Remark. S-BC polynomial and BC-reduction are slightly different from the conventional definitions when using polynomials which have coefficients in a field. However they are natural definitions when working with polynomials which have coefficients in a ring which is not a field, because we want to avoid a division of bracket coefficients.

In both S-BC polynomial and BC-reduction, the resulting coefficients have only two types: *product type*

$$[A, \alpha] \otimes [B, \beta]$$

and *product-difference type*

$$[A, \alpha] \otimes [B, \beta] \ominus [C, \gamma] \otimes [D, \delta].$$

Because in Buchberger's algorithm, any transformation of polynomials is either an S-polynomial calculation or a polynomial reduction, it suffices to consider these two types only as bracket coefficient calculations.

3.2 The Key Theorem

We introduce an algorithm called **R-GB** which we want to mimic. This is a slightly modified version of Buchberger's algorithm. Only the definitions of S-polynomial and reduction are changed so that they correspond to Definitions 5 and 6 for BC polynomials. That is, in **R-GB**, for real polynomials $f = A \cdot LP(f) + rest(f)$ and $g = B \cdot LP(g) + rest(g)$, we define the *S-R polynomial* of f and g by

$$B \cdot \frac{LCM}{LP(f)} \cdot f - A \cdot \frac{LCM}{LP(g)} \cdot g.$$

And for a polynomial f and a finite set F of polynomials,

$f \xrightarrow{R}_F h$ (*R-reduction*) iff there are $g \in F$ and a power product u such that $f = A \cdot u \cdot LP(g) + rest(f)(A \neq 0)$, $g = B \cdot LP(g) + rest(g)$, and $h = B \cdot f - A \cdot u \cdot g$.

It is clear that **R-GB** gives the same result as the usual Buchberger algorithm up to coefficient normalization (i.e. making all the resulting polynomials monic).

Now let E be a (possibly zero) real coefficient of a polynomial[2] at an arbitrary step of **R-GB**.

---

[2] the result or an intermediate polynomial (e.g. $B \cdot \frac{LCM}{LP(f)} \cdot f$ in the above notation) of an S-R polynomial calculation or an R-reduction

Let us consider how E arises. Let $\mathcal{C}$ be the set of all coefficients in the input polynomials. Let $\mathcal{C} = \{A_i\}_{i \in [1,l]}$. For each $i \in [1,l]$, introduce a new indeterminate $\tilde{A}_i$ which corresponds one to one with $A_i \in \mathcal{C}$. Let $\tilde{\mathcal{C}} = \{\tilde{A}_i\}_{i \in [1,l]}$ and $\mathcal{X}$ be the set (formally) generated by $\tilde{\mathcal{C}}$ with two binary relations "$\cdot$" and "$-$". (As the "minus" in usual arithmetic, this "$-$" may also work as a unary operator. That is, if $x \in \mathcal{X}$, then $-x \in \mathcal{X}$.) E at an n-th step ($n \geq 2$) is the result of either $\pm AB$ (product type) where A and B were coefficients at an earlier step or $A - B$ (product-difference type) where $A = ab$ and $B = cd$ where a, b, c, and d were coefficients at an earlier step, because E arises by S-R polynomials and R-reductions. Then from E we can define $\tilde{E} \in \mathcal{X}$ inductively as follows:

$$\tilde{E} = \begin{cases} \tilde{A} \cdot \tilde{B} & \text{if } E \text{ is the result of} \\ (\text{resp. } -\tilde{A} \cdot \tilde{B}) & AB \text{ (resp. } -AB) \\ \tilde{A} - \tilde{B} & \text{if } E \text{ is the result of} \\ & A - B \end{cases}$$

Moreover let $\mathcal{BC}$ be the set of bracket coefficients and their negatives. For a precision μ we define a map $\sigma_\mu : \mathcal{X} \to \mathcal{BC}$ by the following:

$$\sigma_\mu(\tilde{A}_i) = [\langle A_i \rangle_\mu, \alpha_i],$$

where $\langle A_i \rangle_\mu$ is the floating point approximation of A_i to precision μ and α_i is its roundoff error $5 \times 10^{-(\mu+1)} \times 10^{e(\langle A_i \rangle_\mu)}$, and;
For $x, y \in \mathcal{X}$,

$$\sigma_\mu(-x) = -\sigma_\mu(x),$$

$$\sigma_\mu(x \cdot y) = \sigma_\mu(x) \otimes_\mu \sigma_\mu(y),$$

$$\sigma_\mu(x - y) = \sigma_\mu(x) \ominus_\mu \sigma_\mu(y).$$

Recall Definition 4 for BC arithmetic. For E we hereby write

$$[\langle E \rangle_\mu, \epsilon_\mu]$$

as the result of $\sigma_\mu(\tilde{E})$.

For example, if $\tilde{E} = (2\tilde{/}5) \cdot ((1\tilde{/}3) \cdot (3\tilde{/}4) - (5\tilde{/}6) \cdot (1\tilde{/}7)) - (1\tilde{/}2) \cdot (2\tilde{/}3)$, then this means that E arises by $2/5 \times (1/3 \times 3/4 - 5/6 \times 1/7) - 1/2 \times 2/3$, and $\langle E \rangle_5$ is the result of $.40000 \otimes (.33333 \otimes .75000 \ominus .83333 \otimes .14286) \ominus .50000 \otimes .66667$. As to the error, in general, we have the following lemma.

Lemma 1 (Error Analysis) $|\langle E \rangle_\mu - E| \leq \epsilon_\mu$ *for all μ.*

Proof. By induction on the structure of $\tilde{E}$.
(Base case: $\tilde{E} = \tilde{A}_i$) Obvious since $|\langle A_i \rangle_\mu - A_i| \leq \alpha_i = \epsilon_\mu$.
(Multiplication Case: $\tilde{E} = \tilde{A} \cdot \tilde{B}$ or $-\tilde{A} \cdot \tilde{B}$) It suffices to see the case $\tilde{E} = \tilde{A} \cdot \tilde{B}$ only. Let $|\langle A \rangle_\mu - A| \leq \alpha$ and $|\langle B \rangle_\mu - B| \leq \beta$ by induction hypothesis. Let $\natural$ refer to the equation: $|\langle A \rangle_\mu \otimes \langle B \rangle_\mu - AB|$.
When $\langle A \rangle_\mu \otimes \langle B \rangle_\mu \neq 0$:
$\natural$
$= |(\langle A \rangle_\mu - A)(B - \langle B \rangle_\mu) + \langle B \rangle_\mu(\langle A \rangle_\mu - A) + \langle A \rangle_\mu(\langle B \rangle_\mu - B) + (\langle A \rangle_\mu \otimes \langle B \rangle_\mu - \langle A \rangle_\mu \langle B \rangle_\mu)|$
$\leq \alpha\beta + \alpha|\langle B \rangle_\mu| + \beta|\langle A \rangle_\mu| + 5 \times 10^{-(\mu+1)} \times 10^{e(\langle E \rangle_\mu)}$
$\leq \epsilon_\mu$.
When $\langle A \rangle_\mu \otimes \langle B \rangle_\mu = 0$: $\langle A \rangle_\mu \langle B \rangle_\mu = 0$ and so
$\natural = |(\langle A \rangle_\mu - A)(B - \langle B \rangle_\mu) + \langle B \rangle_\mu(\langle A \rangle_\mu - A) + \langle A \rangle_\mu(\langle B \rangle_\mu - B)| \leq \alpha\beta + \alpha|\langle B \rangle_\mu| + \beta|\langle A \rangle_\mu| \leq \epsilon_\mu$
(By Remark after Definition 4).
(Subtraction Case: $\tilde{E} = \tilde{A} - \tilde{B}$) Let $|\langle A \rangle_\mu - A| \leq \alpha$ and $|\langle B \rangle_\mu - B| \leq \beta$ by induction hypothesis. Let $\natural\natural$ refer to the equation: $|(\langle A \rangle_\mu \ominus \langle B \rangle_\mu) - (A - B)|$.
When $\langle A \rangle_\mu \ominus \langle B \rangle_\mu \neq 0$: If $\langle B \rangle_\mu \neq 0$, then there are two cases:
(1) $\langle A \rangle_\mu \ominus \langle B \rangle_\mu \neq \langle A \rangle_\mu$ (normal case), and
(2) $\langle A \rangle_\mu \ominus \langle B \rangle_\mu = \langle A \rangle_\mu$ (loss of information).
In case (1), $\natural\natural = |(\langle A \rangle_\mu - A) + (B - \langle B \rangle_\mu) + \{(\langle A \rangle_\mu \ominus \langle B \rangle_\mu) - (\langle A \rangle_\mu - \langle B \rangle_\mu)\}| \leq \alpha + \beta + 5 \times 10^{-(\mu+1)} \times 10^{e(\langle E \rangle_\mu)} \leq \epsilon_\mu$.
In case (2), it is easy to see from floating point arithmetic that $e(\langle B \rangle_\mu) \leq e(\langle A \rangle_\mu) - \mu$ and when the equality holds, $|\text{mantissa of } \langle B \rangle_\mu| \leq \overbrace{.500\ldots00}^{\mu}$
(i.e. $\leq \overbrace{.500\ldots00}^{\mu}$ if $\langle B \rangle_\mu$ is positive, and
$< \overbrace{.500\ldots00}^{\mu}$ otherwise). Hence when the equality holds, $\langle B \rangle_\mu \leq .5 \times 10^{e(\langle B \rangle_\mu)} = .5 \times 10^{e(\langle A \rangle_\mu) - \mu}$. Then $\natural\natural = |(\langle A \rangle_\mu - A) + (B - \langle B \rangle_\mu) + \langle B \rangle_\mu| \leq \alpha + \beta + |\langle B \rangle_\mu| \leq \alpha + \beta + 5 \times 10^{e(\langle E \rangle_\mu) - (\mu+1)} \leq \epsilon_\mu$.
If the equality does not hold, $\natural\natural \leq \alpha + \beta + |\langle B \rangle_\mu| < \alpha + \beta + 1 \times 10^{e(\langle E \rangle_\mu) - (\mu+1)} < \epsilon_\mu$.
On the other hand, if $\langle B \rangle_\mu = 0$, then $\natural\natural = |(\langle A \rangle_\mu - A) + (B - \langle B \rangle_\mu)| \leq \alpha + \beta < \epsilon_\mu$.
When $\langle A \rangle_\mu \ominus \langle B \rangle_\mu = 0$: we have $\langle A \rangle_\mu - \langle B \rangle_\mu = 0$ from floating point arithmetic, and hence $\natural\natural = |(\langle A \rangle_\mu - A) + (B - \langle B \rangle_\mu)| \leq \alpha + \beta \leq \epsilon_\mu$ ∎

Lemma 2 (Error Convergence) *For the notation as above,*
$$\lim_{\mu \to \infty} \epsilon_\mu = 0.$$

Proof. By induction on the structure of $\tilde{E}$.
(Base case: $\tilde{E} = \tilde{A}_i$) Since $\alpha_i = 5 \times 10^{-(\mu+1)} \times$

$10^{e(\langle A_i\rangle_\mu)}$ and $e(\langle A_i\rangle_\mu)$ is bounded on μ, $\epsilon_\mu = \alpha_i \to 0$.

(Multiplication Case: $\tilde{E} = \tilde{A} \cdot \tilde{B}$ or $-\tilde{A} \cdot \tilde{B}$) It suffices to see the case $\tilde{E} = \tilde{A} \cdot \tilde{B}$ only. Let α and β be the errors of $\langle A\rangle_\mu$ and $\langle B\rangle_\mu$ respectively. When $\langle E\rangle_\mu \neq 0$, $\epsilon_\mu = \uparrow (\uparrow (\alpha\beta) + \uparrow (\alpha|\langle B\rangle_\mu|) + \uparrow (\beta|\langle A\rangle_\mu|) + 5 \times 10^{e((E)_\mu)-(\mu+1)})$. By induction hypothesis, we have $\alpha \to 0$ and $\beta \to 0$. Since $|\langle A\rangle_\mu|$, $|\langle B\rangle_\mu|$, and $e(\langle E\rangle_\mu)$ are bounded on μ, $\epsilon_\mu \to 0$. Similar for the case $\langle E\rangle_\mu = 0$.

(Subtraction case: $\tilde{E} = \tilde{A} - \tilde{B}$) Let α and β be as above. When $\langle E\rangle_\mu \neq 0$, $\epsilon_\mu = \uparrow (\alpha + \beta + 5 \times 10^{e((E)_\mu)-(\mu+1)})$. In the same way, $\epsilon_\mu \to 0$ is implied by that $\alpha \to 0$ and $\beta \to 0$ by induction hypothesis and that $e(\langle E\rangle_\mu)$ is bounded on μ. Similar for the case $\langle E\rangle_\mu = 0$. ∎

Remark. In general, $\{\epsilon_\mu\}_{\mu\geq 1}$ *may not* be monotonically decreasing, as can be seen in the multiplication case, since when $\mu' > \mu$, $|\langle A\rangle_{\mu'}| > |\langle A\rangle_\mu|$ may often occur.

The following is a key theorem for our algorithm.

Theorem 1 (Zero Judgment) *Let E be a real coefficient of a polynomial at any step of* **R-GB**. *Then*

$$E = 0 \quad iff \quad |\langle E\rangle_\mu| \leq \epsilon_\mu \text{ for all } \mu.$$

Remark. The following also holds:
"Let E be a real coefficient *of product-difference type* of a polynomial at any step of **R-GB**. Then $E = 0$ iff $|\langle E\rangle_\mu| \leq \epsilon_\mu$ for all μ."
In fact, if E is of product type, E cannot be zero by the structure of **R-GB**.
Proof. ($\Rightarrow$) Immediate from Lemma 1.
($\Leftarrow$) For any μ, $|E| \leq |E - \langle E\rangle_\mu| + |\langle E\rangle_\mu|$. $|E - \langle E\rangle_\mu| \leq \epsilon_\mu$ by Lemma 1 and $|\langle E\rangle_\mu| \leq \epsilon_\mu$ by assumption. Thus $|E| \leq 2\epsilon_\mu$ for all μ. But by Lemma 2, $\epsilon_\mu \to 0$ as μ approaches infinity. Therefore E must be 0. ∎

4 The Algorithm

4.1 Description

For simplicity, we describe the algorithm **FP-GB** that computes an approximate Gröbner basis, based on the crude version of Buchberger's algorithm ([3], Algorithm 6.2). Thus here **R-GB** is the algorithm obtained by replacing S-polynomials and reductions in Algorithm 6.2 by S-R polynomials and R-reductions respectively.

In the light of Theorem 1, we put a basic assumption for the zero judgment in **FP-GB** as follows:
For any bracket coefficient $[A, \alpha]$ such that A is product-difference type,

Assumption Z: $\quad [A, \alpha] = 0 \quad$ iff $\quad |A| \leq \alpha$.

This assumption will be shown to be adequate for our purpose in the correctness (Section 4.2) of **FP-GB**. Now let us describe **FP-GB**.

An admissible term ordering $<_T$ is given.

Algorithm FP-GB
Input: a finite set F of real polynomials and a natural number μ.
Output: a set G_μ of floating point polynomials with precision μ, such that $\{G_\mu\}_\mu$ makes a floating point Gröbner sequence of F.

$G :=$ **R-to-BC** (F, μ) % Data conversion to initialize
$B := \{\{f_1, f_2\} \mid f_1, f_2 \in G, f_1 \neq f_2\}$
While $B \neq \emptyset$ do
    $(f_1, f_2) :=$ a pair in B
    $B := B - \{\{f_1, f_2\}\}$
    $h :=$ **S-BCpoly** (f_1, f_2, μ)
    $h' :=$ **BC-NForm** (h, G, μ)
    if $h' \neq 0$ then
        if $h' =$ a bracket without the
        variables then Return $(\{1\})$ else
        $B := B \cup \{\{g, h'\} \mid g \in G\}$
        $G := G \cup \{h'\}$
BC-to-FP(G) % Data conversion to finish
                % If necessary, the final errors
                can be viewed from G.

Subalgorithm R-to-BC (F, μ) % F is a finite set of real polynomials

$BCF := \emptyset$
For f in F do
    $n :=$ the number of the terms in f
    $BCf := 0$
    For $i = 1$ to n do
        $A_i :=$ the i-th coefficient in f
        $T_i :=$ the i-th power product in f
        $\langle A_i\rangle_\mu :=$ the floating point approximation of A_i to precision μ
        $\alpha_i := 5 \times 10^{-(\mu+1)} \times 10^{e(\langle A_i\rangle_\mu)}$ (∗)
        $BCf := BCf + [\langle A_i\rangle_\mu, \alpha_i]T_i$
    $BCF := BCF \cup \{BCf\}$

Remark (∗): If an input coefficient A_i is initially given by a floating point number with precision μ, then we can set $\alpha_i := 0$ for it.

Subalgorithm S-BCpoly $(f_1,\ f_2,\ \mu)$ % f_1 and f_2 are BC polynomials with precision μ

$f := \text{S-BCpoly}_\mu(f_1, f_2)$
$n :=$ the number of the terms in f
$fZ := 0$
For $i = 1$ to n do
  $[E^i_\mu, \epsilon^i_\mu] :=$ the i-th bracket coefficient in f
  $T_i :=$ the i-th power product in f
  if E^i_μ is of product-difference type and
  $|E^i_\mu| \leq \epsilon^i_\mu$
  then $fZ := fZ$ % **Assumption Z**
  else $fZ := fZ + [E^i_\mu, \epsilon^i_\mu]T_i$

Subalgorithm BC-NForm $(f,\ G,\ \mu)$ % f is a BC polynomial and G is a finite set of BC polynomials with precision μ

$h := f$
While $\exists g \in G, u$ such that $h \xrightarrow[\ g,u\ ]{BC}$ do
  choose $g \in G, u$ such that $h \xrightarrow[\ g,u\ ]{BC}$
  and $u \cdot LP(g)$ is maximal (w.r.t. $<_T$)
  $[A, \alpha] :=$ the bracket coefficient of $LP(g)$ in h
  $[B, \beta] :=$ the leading bracket coefficient in g
  $h := [B, \beta] \cdot h - [A, \alpha]u \cdot g$ (with precision μ)
  $n :=$ the number of the terms in h
  $hZ := 0$
  For $i = 1$ to n do
    $[E^i_\mu, \epsilon^i_\mu] :=$ the i-th bracket coefficient in h
    $T_i :=$ the i-th power product in h
    if E^i_μ is of product-difference type and
    $|E^i_\mu| \leq \epsilon^i_\mu$
    then $hZ := hZ$ % **Assumption Z**
    else $hZ := hZ + [E^i_\mu, \epsilon^i_\mu]T_i$
  $h := hZ$

Subalgorithm BC-to-FP (G) % G is a finite set of BC polynomials

$FPG := \emptyset$
For g in G do
  $n :=$ the number of the terms in g
  $FPg := 0$
  For $i = 1$ to n do
    $[E_i, \epsilon_i] :=$ the i-th coefficient in g
    $T_i :=$ the i-th power product in g
    $FPg := FPg + E_i \cdot T_i$
  $FPG := FPG \cup \{FPg\}$

4.2   Termination and Correctness

The termination of **FP-GB** can be shown in exactly the same manner as Buchberger's proof ([4]) for the conventional algorithm using Dickson's lemma ([6]), because the termination depends only on a property of a sequence of *power products*, not on *coefficients*.

Theorem 2 (Termination) *For any finite set F of real polynomials and any natural number μ, **FP-GB** (F, μ) terminates.*

Proof. Let t_i be the leading power product of i-th BC polynomial h_i adjoined to G in the course of **FP-GB** $(i = 1, 2, \ldots)$. Assume that $\exists j$ such that t_j is a multiple of t_k for some $k < j$. Then h_j is reducible using h_k. However, by the structure of **FP-GB**, h_j is a normal form modulo the $(j-1)$-th set G including h_k. This is a contradiction. Thus the sequence $t_1, t_2, \ldots$ has a property that, for all j, t_j is not a multiple of any of its predecessors. Hence, by Dickson's lemma, this sequence must be finite. ∎

The correctness of **FP-GB** is the following.

Theorem 3 (Correctness) *Given a finite set F of real polynomials and a natural number μ, let $G_\mu = $ **FP-GB** (F, μ). Then $\{G_\mu\}_\mu$ is a floating point Gröbner sequence of F.*

Proof. By comparison to **R-GB**. Let G be the result of **R-GB** (F). First of all, we prove that there is an M such that $\mathcal{M}(G_\mu) = \mathcal{M}(G)$ for all $\mu \geq M$. As mentioned in Section 3, it is obvious that if at any stage of **R-GB**, a coefficient E is zero iff **FP-GB** judges that the corresponding $[\langle E \rangle_\mu, \epsilon_\mu]$ is zero, then $\mathcal{M}(G_\mu) = \mathcal{M}(G)$. Here we could consider four cases:

	R-GB	FP-GB
(1)	$E = 0$	$[\langle E \rangle_\mu, \epsilon_\mu] = 0$
(2)	$E = 0$	$[\langle E \rangle_\mu, \epsilon_\mu] \neq 0$
(3)	$E \neq 0$	$[\langle E \rangle_\mu, \epsilon_\mu] = 0$
(4)	$E \neq 0$	$[\langle E \rangle_\mu, \epsilon_\mu] \neq 0$

Cases (1) and (4) are desired for our purpose. Case (2) is impossible by Theorem 1. However, the undesirable case (3) is possible. Therefore it suffices to prove that there is an M such that for all $\mu \geq M$ case (3) does not occur throughout **FP-GB** (F, μ).

Let N be the sum of the numbers of S-R polynomial calculations and R-reductions in **R-GB** (F). Note that N is finite. Here when $N = 0$, it means $\sharp F = 1$, and hence we are done by **FP-GB** $(F, 1)$, i.e. taking 1 as M. Thus we assume $N \geq 1$.

Let $\{s_1, s_2, \ldots, s_N\}$ be the sequence of S-R polynomial calculations and R-reductions such that they are serially performed in $\mathbf{R\text{-}GB}\,(F)$ in this order. Put $\mathcal{S}_k = \{s_1, \ldots, s_k\}$ for $1 \le k \le N$.

We prepare the following lemma[3].

$(*)$ For any $k \in [1, N]$, if for some M_0, $\mathbf{FP\text{-}GB}\,(F, M_0)$ causes no case (3) for $\mathcal{S}_k$, then there is an $M'(\ge M_0)$ such that for all $\mu \ge M'$, $\mathbf{FP\text{-}GB}\,(F, \mu)$ causes no case (3) for $\mathcal{S}_k$.

Proof of $(*)$. Assume the if part. If moreover no case (4) occurs in $\mathcal{S}_k$, then it means $E = 0$ for every coefficient E and hence no case (3) can occur for any μ. Otherwise, let $\{E_i, [\langle E_i \rangle_\mu, \epsilon_\mu^i]\}_{1 \le i \le l}$ be all the pairs in case (4) in $\mathcal{S}_k$. Put $C = \min_{i \in [1, l]}(|E_i|)$. Since $C \ne 0$ and l is finite, there is an M_1 such that $\mu \ge M_1 \Rightarrow \epsilon_\mu^i < C/2$ for all i, by Lemma 2. For $i \in [1, l]$ and $\mu \ge M_1$, we have $\epsilon_\mu^i \ge |E_i - \langle E_i \rangle_\mu| \ge ||E_i| - |\langle E_i \rangle_\mu||$ by Lemma 1 and the well-known inequality about absolute values. Thus when $|\langle E_i \rangle_\mu| \le |E_i|$, $|\langle E_i \rangle_\mu| \ge |E_i| - \epsilon_\mu^i \ge C - \epsilon_\mu^i > \epsilon_\mu^i$ and when $|\langle E_i \rangle_\mu| > |E_i|$, $|\langle E_i \rangle_\mu| > |E_i| \ge C > 2\epsilon_\mu^i > \epsilon_\mu^i$. Therefore, $M' = \max(M_0, M_1)$ satisfies the subject of $(*)$ by **Assumption Z**. (*End of Proof of* $(*)$)

Now if $\mathbf{FP\text{-}GB}\,(F, 1)$ causes no case (3) for $\mathcal{S}_N$, then we are done by $(*)$. Otherwise, let s_{k_1} be the first process such that case (3) occurs. Let $E_1, \ldots E_m$ be all the coefficients in case (3) immediately after the process s_{k_1}. By an argument similar to the proof of $(*)$, there is a μ_{k_1} such that $\mu \ge \mu_{k_1} \Rightarrow |\langle E_i \rangle_\mu| > \epsilon_\mu^i$ for all $i \in [1, m]$. Moreover, since $\mathbf{FP\text{-}GB}\,(F, 1)$ causes no case (3) for $\mathcal{S}_{k-1}$, by $(*)$ there is an M_{k_1} such that for all $\mu \ge M_{k_1}$, $\mathbf{FP\text{-}GB}\,(F, \mu)$ causes no case (3) for $\mathcal{S}_{k_1 - 1}$. Therefore, when putting $\tilde{M}_{k_1} = \max(\mu_{k_1}, M_{k_1})$, $\mathbf{FP\text{-}GB}\,(F, \tilde{M}_{k_1})$ causes no case (3) for $\mathcal{S}_{k_1}$.

Next if $\mathbf{FP\text{-}GB}\,(F, \tilde{M}_{k_1})$ causes no case (3) for $\mathcal{S}_N$, then we are done. Otherwise, let s_{k_2} be the first process such that case (3) occurs. Repeating this argument makes a sequence $k_1 < k_2 < k_3 < \ldots$. But since N is finite, this sequence is finite and hence it proves the existence of M.

Moreover, as μ approaches infinity after M, by Lemmas 1 and 2, every coefficient of $\{G_\mu\}_{\mu \ge M}$ converges to the corresponding one of G. ∎

5 Examples

We can also consider an improved version of $\mathbf{FP\text{-}GB}$, based on the improved version of Buchberger's algorithm ([3], Algorithm 6.3), in a man-

ner similar to $\mathbf{FP\text{-}GB}$. In fact, the optimum choice of a pair for S-polynomial, and Criterions 1 and 2 for detecting unnecessary reductions are all expressed only by the terminology of power products, not of coefficients. Obviously the termination and correctness of the improved $\mathbf{FP\text{-}GB}$ can be proved in the same manner as those of $\mathbf{FP\text{-}GB}$ in Section 4.2.

The original versions of $\mathbf{FP\text{-}GB}$ and $\mathbf{R\text{-}GB}$ are so costly in computing time that they are not appropriate for our experiments. Thus, in this section, we compare the experimental results of the improved $\mathbf{FP\text{-}GB}$ and the improved $\mathbf{R\text{-}GB}$, where the latter algorithm gives a *reduced Gröbner basis* except that the resulting polynomials are not necessarily monic. Let us here call them simply $\mathbf{FP\text{-}GB}$ and $\mathbf{R\text{-}GB}$ as well.

We implemented them in Maple V/Sun SPARC ([5]). For brevity and convenience on Maple, we coded the parts of error terms in Definition 4 by floating point arithmetic instead of round up arithmetic, assuming that the errors between them are negligible. We did not use the built-in function *gbasis*[4] in Maple for computing Gröbner basis even in the case of rationals, for the sake of fairness of the comparison between $\mathbf{FP\text{-}GB}$ and $\mathbf{R\text{-}GB}$. Tables 1 to 5 show $G := \mathbf{R\text{-}GB}\,(F)$ and $G_\mu := \mathbf{FP\text{-}GB}\,(F, \mu)$ $(1 \le \mu \le 10 \text{ or } 20)$ with cpu times for F in Examples 1 to 5 below, respectively. To preserve space, in Tables 3 and 4, each coefficient of G is expressed by floating point with precision 10. *tdeg* (resp. *plex*) denotes total degree lexicographic (resp. purely lexicographic) term ordering. In the tables except for Table 5, the first precision $\tilde{M}$ satisfying $\mathcal{M}(G_{\tilde{M}}) = \mathcal{M}(G)$ is marked in bold. Note that $\tilde{M}$ may not necessarily satisfy the condition that $\mathcal{M}(G_\mu) = \mathcal{M}(G)$ *for all* $\mu \ge \tilde{M}$. In Table 5, the precision such that from it on, a series of monomial supports *seems* to be stable is framed by a box. Here are the examples:

1. $F = \{x(3x - 1), x - 1/3\}$, tdeg.

2. $F = \{x(3x - 1), x - .3333333333\}$, tdeg.

3. $F = \{f_1, f_2, f_3\}$, plex with $x > y > z$, where
$f_1 = \frac{1}{7}x^2 - \frac{326548390854652}{272974017239}x + \frac{1263781236281}{712638126}y^2 + \frac{26872672361827}{7263188218281}z^2$, $f_2 = \frac{3}{8}xy + \frac{12367812638123}{763812368213132}yz - \frac{63812638126}{77263812831}y$, $f_3 = \frac{4}{9}x + \frac{327091270979304}{24122375460421}y + \frac{18467031595309203}{318405459032}z - \frac{35631806369314319}{6436561806418109}$.

---

[3] We need this lemma, since for an E, $\{\epsilon_\mu\}_\mu$ may not be monotonically decreasing as mentioned in Remark after Lemma 2.

[4] This function can be applied only over the rationals, but it is much faster than $\mathbf{R\text{-}GB}$.

4. $F = \{\sqrt{2}ex^3y + \sqrt{3}xy + \frac{\sqrt{7}}{e}, \frac{\sqrt{3}}{e}x^2y^2 - \sqrt{7}xy + \frac{e\sqrt{11}}{11}\}$, tdeg with $x > y$, where e is Napier's number ($2.71828\ldots$).

5. $F = \{f_1, f_2\}$, tdeg with $x > y$, where
$f_1 = 2\sqrt{2}/\pi x^3y + (\sqrt{3} + \pi)xy + \sqrt{7}/(e - \pi)$,
$f_2 = (1 - e\sqrt{3})/e \cdot \pi x^2y^2 - (\sqrt{7} - e)xy + e/\sqrt{11}$.

Remarks. 1. In Example 3, the reason why only the leading coefficients have small numerators and denominators is that $\tilde{M}(= 6)$ was small. According to our experiments, $\tilde{M}$ was 14 when they were all replaced by a rational number with numerator and denominator of about 15 digits. (It took 261.150 sec. for $\tilde{M} = 14$, whereas **R-GB** required 1278.63 sec.) In the case of rational field, in general, we have a conjecture that $\tilde{M}$ depends largely on the sizes of the numerators and denominators of the leading coefficients of input polynomials.
2. Our approach could be useless if input polynomials contain a polynomial of too high degree or with too many terms, because even using floating point computation does not solve the problem of the number of reductions or the number of polynomials which we must retain to make S-polynomials. In fact, it was not possible to compute a lexicographic floating point Gröbner basis of the "Rose system" (see [2]) which has a polynomial with 21 terms, using **FP-GB** within a reasonable time. Our approach will be useful when *the growth of coefficients* is the main reason why the computation of a Gröbner basis by the usual Buchberger algorithm is slow (see Examples 3,4,5).

6 Conclusion

Our algorithm **FP-GB** turns out to be more efficient than the conventional **R-GB**, in particular when the coefficients of input data are complicated, such as real numbers including irrational or transcendental numbers, and the growth of intermediate coefficients is the major factor determining the computational cost. The algorithm can be easily extented to complex numbers as well.

Unfortunately, up to now, we do not have a reasonable upper bound on M such that $\mathcal{M}(G_\mu) = \mathcal{M}(G)$ for all $\mu \geq M$. That is, we cannot strictly determine M in advance where G_M is an approximate Gröbner basis. In this sense, the algorithm is probablistic. In practice, currently, one should guess M by detecting a subsequence of $\{G_\mu\}_\mu$ that

seems to be stable on monomial support or seems to converge coefficientwise.

We want to stress, however, that our approach is not only theoretically well-founded, but also could be practically useful, in particular for approximate algebraic computation in multivariate polynomials.

Acknowledgment

This research is supported in part by the United States Army Research Office through the Army Center of Excellence for Symbolic Methods in Algorithmic Mathematics (ACSyAM), Mathematical Sciences Institute of Cornell University, Contract DAAL03-91-C-0027.

The author is grateful to Moss Sweedler for his helpful discussion and valuable comments in preparing this paper. Also the author wishes to thank James Davenport, Jerry Marsden, and Dana Scott for their suggestion of some relationship to interval arithmetic, and Tateaki Sasaki and Matu-Tarow Noda for their encouragement.

References

[1] Alefeld, G. and Herzberger, J., *Introduction to Interval Computations, Computer Science and Applied Mathematics, Academic Press* (1983).

[2] Boege, W., Gebauer, R., and Kredel, H., Some Examples for Solving Systems of Algebraic Equations by Calculating Groebner Bases, *J. Symb. Comp.* **1** (1986), 83-98.

[3] Buchberger, B., Gröbner Bases: An Algorithmic Method in Polynomial Ideal Theory, *Chapter 6 in Multidimensional Systems Theory (N. K. Bose ed.), D. Reidel Publishing Company* (1985), 184-232.

[4] Buchberger, B., An algorithmical criterion for the solvability of algebraic systems of equations (German), *Aequationes mathematicae* **4**(3) (1970), 374-383.

[5] Char, B. W., Geddes, K. O., Gonnet, G. H., Leong, B. L., Monagan, M. B., and Watt, S. M., *First Leaves: A Tutorial Introduction to Maple V, Springer-Verlag* (1992).

[6] Dickson, L. E., Finiteness of the odd perfect and primitive abundant numbers with n

Table 1: Example 1

	G	cpu time (sec)
	$\{x - 1/3\}$	.130

μ	G_μ	cpu time (sec)
1	$\{1.x - .3\}$	.233
2	$\{1.x - .33\}$	.167
3	$\{1.x - .333\}$	.200
4	$\{1.x - .3333\}$	.184
5	$\{1.x - .33333\}$	.184
6	$\{1.x - .333333\}$	.183
7	$\{1.x - .3333333\}$	.167
8	$\{1.x - .33333333\}$	.150
9	$\{1.x - .333333333\}$	.167
10	$\{1.x - .3333333333\}$	.183
11	$\{1.x - .33333333333\}$	.200
12	$\{1.x - .333333333333\}$	.183
13	$\{1.x - .3333333333333\}$	.200
14	$\{1.x - .33333333333333\}$	.200
15	$\{1.x - .333333333333333\}$	.183
16	$\{1.x - .3333333333333333\}$	.200
17	$\{1.x - .33333333333333333\}$	.216
18	$\{1.x - .333333333333333333\}$	.184
19	$\{1.x - .3333333333333333333\}$	.200
20	$\{1.x - .33333333333333333333\}$	.200

Note. G_μ gives an *approximate GCD* of the input polynomials.

Table 2: Example 2

	G	cpu time (sec)
	$\{1\}$	.200

μ	G_μ	cpu time (sec)
1	$\{1.x - .3\}$	.150
2	$\{1.x - .33\}$	.183
3	$\{1.x - .333\}$	.134
4	$\{1.x - .3333\}$	.134
5	$\{1.x - .33333\}$	.116
6	$\{1.x - .333333\}$	.150
7	$\{1.x - .3333333\}$	.134
8	$\{1.x - .33333333\}$	.150
9	$\{1.x - .333333333\}$	.150
10	$\{1.x - .3333333333\}$	.150
11	$\{1.x - .3333333333\}$	.150
12	$\{1\}$	.217
13	$\{1\}$	.200
14	$\{1\}$	.217
15	$\{1\}$	.216
16	$\{1\}$	.233
17	$\{1\}$	.217
18	$\{1\}$	.217
19	$\{1\}$	.200
20	$\{1\}$	.200

Table 3: Example 3

floating point expression of G with precision 10	cpu time (sec)
$\{-.7486148880 \times 10^{31}\boldsymbol{x} + .2938720185 \times 10^{40}\boldsymbol{z^3} + .1954874591 \times 10^{39}\boldsymbol{z^2} - .3745685851 \times 10^{36}\boldsymbol{z} + .1768674726 \times 10^{33},$ $-.1684383498 \times 10^{32}\boldsymbol{y} - .2167251711 \times 10^{39}\boldsymbol{z^3} - .1441683806 \times 10^{38}\boldsymbol{z^2} - .4442206312 \times 10^{35}\boldsymbol{z} + .5572270445 \times 10^{32},$ $-.5081835293 \times 10^{76}\boldsymbol{z^4} - .3068656908 \times 10^{75}\boldsymbol{z^3} + .8958744250 \times 10^{72}\boldsymbol{z^2} - .8274384887 \times 10^{69}\boldsymbol{z} + .2446596580 \times 10^{66}\}$	369.16

μ	G_μ	cpu time (sec)
1	$\{-.4 \times 10^{11}z^2, -8.x + 1000., -20.y - 100000.z\}$	1.833
2	$\{-.46 \times 10^{168}x - .48 \times 10^{173}z + .61 \times 10^{171}, -.10 \times 10^{169}y, .24 \times 10^{152}z^2 + .14 \times 10^{151}z - .18 \times 10\}$	12.000
3	$\{-.349 \times 10^{126}z^2, .136 \times 10^{143}y + .478 \times 10^{146}z - .464 \times 10^{143}, .607 \times 10^{142}x\}$	8.817
4	$\{.3025 \times 10^{1576}y + .7964 \times 10^{1579}z - .1002 \times 10^{1577},$ $.1344 \times 10^{1576}x + .6750 \times 10^{1580}z - .3169 \times 10^{1577}, -.9180 \times 10^{771}z^2\}$	102.600
5	$\{-.11300 \times 10^{2591}y + .37380 \times 10^{2591}, .12153 \times 10^{1106}z, -.50220 \times 10^{2590}x + .11864 \times 10^{2592}\}$	128.216
6	$\{.537759 \times 10^{25}\boldsymbol{x} - .211101 \times 10^{34}\boldsymbol{z^3} - .140427 \times 10^{33}\boldsymbol{z^2} + .269066 \times 10^{30}\boldsymbol{z} - .127046 \times 10^{27},$ $.120996 \times 10^{26}\boldsymbol{y} + .155683 \times 10^{33}\boldsymbol{z^3} + .103562 \times 10^{32}\boldsymbol{z^2} + .319102 \times 10^{29}\boldsymbol{z} - .400280 \times 10^{26},$ $-.262230 \times 10^{64}\boldsymbol{z^4} - .158347 \times 10^{63}\boldsymbol{z^3} + .462286 \times 10^{60}\boldsymbol{z^2} - .42695 \times 10^{57}\boldsymbol{z} + .12621 \times 10^{54}\}$	24.650
7	$\{-.5377591 \times 10^{25}x + .2110991 \times 10^{34}z^3 + .1404259 \times 10^{33}z^2 - .2690671 \times 10^{30}z + .1270510 \times 10^{27},$ $-.1209958 \times 10^{26}y - .1556817 \times 10^{33}z^3 - .1035615 \times 10^{32}z^2 - .3191010 \times 10^{29}z + .4002777 \times 10^{26},$ $-.2622272 \times 10^{64}z^4 - .1583455 \times 10^{63}z^3 + .4622788 \times 10^{60}z^2 - .426967 \times 10^{57}z + .126248 \times 10^{54}\}$	24.567
8	$\{-.74861510 \times 10^{31}x + .29387212 \times 10^{40}z^3 + .19548753 \times 10^{39}z^2 - .37456868 \times 10^{36}z + .17686754 \times 10^{33},$ $-.16843840 \times 10^{32}y - .21672524 \times 10^{39}z^3 - .14416843 \times 10^{38}z^2 - .44422077 \times 10^{35}z + .55722718 \times 10^{32},$ $-.50818387 \times 10^{76}z^4 - .30686591 \times 10^{75}z^3 + .8958751 \times 10^{72}z^2 - .8274391 \times 10^{69}z + .2446602 \times 10^{66}\}$	30.100
9	$\{.115073820 \times 10^{28}x - .451727265 \times 10^{36}z^3 - .300494807 \times 10^{35}z^2 + .575770523 \times 10^{32}z - .27187297 \times 10^{29},$ $.258916096 \times 10^{28}y + .333140493 \times 10^{35}z^3 + .221609356 \times 10^{34}z^2 + .68283661 \times 10^{31}z - .85654515 \times 10^{28},$ $.120076142 \times 10^{69}z^4 + .725077583 \times 10^{67}z^3 - .211681686 \times 10^{65}z^2 + .19551130 \times 10^{62}z - .5780940 \times 10^{58}\}$	27.483
10	$\{.5377596431 \times 10^{25}x - .2110998780 \times 10^{34}z^3 - .1404263631 \times 10^{33}z^2 + .2690674097 \times 10^{30}z - .1270508920 \times 10^{27},$ $.1209959197 \times 10^{26}y + .1556822505 \times 10^{33}z^3 + .1035618421 \times 10^{32}z^2 + .3191012254 \times 10^{29}z - .4002781954 \times 10^{26},$ $-.2622286133 \times 10^{64}z^4 - .1583462665 \times 10^{63}z^3 + .4622816264 \times 10^{60}z^2 - .426967888 \times 10^{57}z + .126247231 \times 10^{54}\}$	24.617

Note. Even for $\mu = 100$, it took only 35.250 sec.

Table 4: Example 4

floating point expression of G with precision 10	cpu time (sec)
$\{-.1325528855 \times 10^9\boldsymbol{y^3} - .1085028162 \times 10^{10}\boldsymbol{x} - .3879281551 \times 10^{10}\boldsymbol{y},$ $.3443764079 \times 10^9\boldsymbol{x^2} + .8401300998 \times 10^7\boldsymbol{y^2} + .4145701265 \times 10^9,$ $.3975253926 \times 10^7\boldsymbol{xy} + 492680.0344\boldsymbol{y^2} - .1293658022 \times 10^7\}$	43.21

μ	G_μ	cpu time (sec)
1	$\{-80.x + 30.y, 7000.y^4 + 20000.y^2 + 500000.\}$	2.267
2	$\{83.x^2 + 100., -2400.xy - 690.y^2 + 730., -63000.y^3 - 620000.x\}$	2.567
3	$\{1\}$	2.900
4	$\{1\}$	2.700
5	$\{-229890.\boldsymbol{y^3} - .18815 \times 10^7\boldsymbol{x} - .67274 \times 10^7\boldsymbol{y},$ $-597150.\boldsymbol{x^2} - 14570.\boldsymbol{y^2} - 718900.,$ $-6892.6\boldsymbol{xy} - 854.35\boldsymbol{y^2} + 2243.1\}$	3.550
6	$\{-.132550 \times 10^9 y^3 - .108499 \times 10^{10}x - .387913 \times 10^{10}y,$ $.344366 \times 10^9 x^2 + .840125 \times 10^7 y^2 + .414559 \times 10^9,$ $.397519 \times 10^7 xy + 492664.y^2 - .129365 \times 10^7\}$	4.583
7	$\{-.1325529 \times 10^9 y^3 - .1085029 \times 10^{10}x - .3879282 \times 10^{10}y,$ $.3443766 \times 10^9 x^2 + .8401299 \times 10^7 y^2 + .4145703 \times 10^9,$ $.3975255 \times 10^7 xy + 492680.1 y^2 - .1293659 \times 10^7\}$	4.983
8	$\{-229800.22 y^3 - .18810585 \times 10^7 x - .67253136 \times 10^7 y,$ $-597027.96 x^2 - 14564.909 y^2 - 718719.25,$ $-6891.6968 xy - 854.13454 y^2 + 2242.7495\}$	3.566
9	$\{-.132552891 \times 10^9 y^3 - .108502819 \times 10^{10}x - .387928170 \times 10^{10}y,$ $.344376416 \times 10^9 x^2 + .840130139 \times 10^7 y^2 + .414570140 \times 10^9,$ $.397525401 \times 10^7 xy + 492680.049 y^2 - .129365806 \times 10^7\}$	4.583
10	$\{-.1325528856 \times 10^9 y^3 - .1085028160 \times 10^{10}x - .3879281549 \times 10^{10}y,$ $.3443764073 \times 10^9 x^2 + .8401301011 \times 10^7 y^2 + .4145701258 \times 10^9,$ $.3975253922 \times 10^7 xy + 492680.0346 y^2 - .1293658021 \times 10^7\}$	4.967

Note. Even for $\mu = 100$, it took only 8.517 sec.

Table 5: Example 5

G	cpu time (sec)
?	$> 3,600.$

μ	G_μ	cpu time (sec)
2	$\{-12.x^2 - 55., 48000.xy + 160000.y^2 + 7700., -.38 \times 10^8 y^3 + 580000.x\}$	3.200
3	$\{1\}$	1.550
4	$\{-.2929 \times 10^{12}x, .9853 \times 10^{19}y^2 - .1421 \times 10^{18}\}$	3.883
5	$\{1\}$	3.133
6	$\{-.304467 \times 10^8 y^3 - 57296.6x + .177025 \times 10^7 y,$ $57131.5x^2 - .349660 \times 10^8 y^2 + 503082.,$ $-5860.01xy + 134278.y^2 - 958.949\}$	4.367
7	$\{523698.7y^3 + 985.5738x - 30450.19y,$ $982.7229x^2 - 601411.7y^2 + 8653.376,$ $-100.7948xy + 2309.56y^2 - 16.49430\}$	3.883
8	$\{-.30449275 \times 10^8 y^3 - 57303.400x + .17704401 \times 10^7 y,$ $57137.690x^2 - .34967823 \times 10^8 y^2 + 503126.47,$ $-5860.4880xy + 134285.50y^2 - 959.02183\}$	4.750
9	$\{-.304493527 \times 10^8 y^3 - 57303.5247x + .177044409 \times 10^7 y,$ $57137.8102x^2 - .349679197 \times 10^8 y^2 + 503127.527,$ $-5860.49750xy + 134285.767y^2 - 959.023145\}$	4.767
10	$\{-.3044934742 \times 10^8 y^3 - 57303.51814x + .1770443855 \times 10^7 y,$ $57137.80337x^2 - .3496791346 \times 10^8 y^2 + 503127.4531,$ $-5860.497001xy + 134285.7486y^2 - 959.0230326\}$	4.316

Notes.   1. **R-GB**(F) was not obtained after 3,600 sec.
2. For $\mu = 1$, the constant term of f_1 is ill-defined since $e \sim \pi$.
3. Even for $\mu = 100$, it took only 7.950 sec.

distinct prime factors, *Am. J. of Math.* **35** (1913), 413-426.

[7] Knuth, D. E., *The Art of Computer Programming, Vol. 2, Addison-Wesley* (1969).

[8] Moore, R. E., Interval Arithmetic and Automatic Error Analysis in Digital Computing, *Ph.D. Thesis, Mathematics Dept., Stanford Univ.*, October (1962).

[9] Petković, M., *Iterative Methods for Simultaneous Inclusion of Polynomial Zeros, L. N. Math., Springer-Verlag* **1387** (1989).

[10] Sasaki, T. and Takeshima, T., A Modular Method for Gröbner-basis Construction over Q and Solving System of Algebraic Equations, *J. Information Processing* **12**(4) (1989), 371-379.

[11] Trinks, W., On Improving Approximate Results of Buchberger's Algorithm by Newton's Method, *SIGSAM Bull.* **18**(3) (1984), 7-11.

[12] Winkler, F., A *p*-adic Approach to the Computation of Gröbner Bases, *J. Symb. Comp.* **6**/2,3 (1988), 287-304.

Brief Biography of the Author.
Date and Place of Birth: 10 December 1957, Shizuoka, Japan. Degrees: B.Sc. (Math.) 1982, M.Sc. (Math.) 1984, Ph.D. (Math. Sc.) 1992, University of Tokyo. Current Employment: Senior Research Scientist of Nippon Telegraph and Telephone Corporation (NTT), Visiting Scholar of MSI, Cornell University (1992-1993). Research Interests: Computational algebra, Computer Go, Machine Learning. Academic Societies: Editorial Board of Japan Society for Symbolic and Algebraic Computation. Address: NTT Communication Science Laboratories, Hikaridai, Seika-cho, Soraku-gun, Kyoto 619-02 Japan   (Email: shirayan@progn.kecl.ntt.jp).

III

MAPLE V IN SCIENCE AND ENGINEERING

Part A: Modeling and Simulation

THE USE OF MAPLE FOR MULTIBODY SYSTEMS MODELING AND SIMULATION

P. Capolsini
Projet SAFIR, INRIA - Université de Nice Sophia-Antipolis, France

Abstract

Conventional numerical general-purpose mechanical simulation programs require more time for performing simulations than do special-purpose "symbolic" ones. In addition, these softwares, although said to be "symbolic" fail to provide explicit motion equations of the considered system. It is shown, in this paper, that the use of a general computer algebra system like Maple can be used to write a new complete and efficient symbolic program which presents both advantages to provide explicit equations as well as excellent run-time numerical simulations.

1 Introduction

This paper intends to deal with multibody systems' dynamical equations derivation. The problem is rather old but still has a modern flavor with the recent needs of the spacecraft industry. Handwriting such equations, even for a modestly complex system, is a tedious, long and expensive work and leads to equations subject to careless errors. Accordingly, much attention has been directed toward the development of computer programs to solve the problem. Oldest attempts at writing softwares (such as ADAMS [1]) consisted in establishing the equations of the considered mechanism by specializing a set of generic equations derived from a large and general system.

Mathematical Computation with Maple V:
Ideas and Applications
Tom Lee, Editor
©1993 Birkhäuser Boston

While these general-purpose multibody programs are still being widely employed they can become inadequate to provide efficient simulation code for a given system. A newer approach consists in deriving, directly from the description of the mechanism, a specialized numerical simulation code. This method leads directly to the equations of the considered mechanism and is therefore much more efficient.

Indeed, the newest widespread method consists in writing new "symbolic" special-purpose simulation programs. Let's cite, among others, AUTOLEV [2], DYNAMICA [3], MESA-VERDE [4], SD-EXACT, SD-FAST [5] and so on ...These softwares are all said to be "symbolic" because the user is not compelled to give explicit numerical values to all the indeterminates and they use some typical symbolic algebra methods. Although these new programs supply efficient simulation codes, they still fail in providing explicitly the set of equations governing the behavior of the mechanism. Let us notice that these equations may be so interesting to analysts that a few of them still go on deriving motion equations by hand !

During the last decade, the arrival of computer algebra has offered a new mean to supply explicit equations as well as simulation codes.

2 Symbolic architecture

All aforementioned "symbolic" simulation softwares share, with other programs like our own prototype or JAMES [6], approximately the same architecture which is described in figure 1. The greatest difference between these two families of symbolic programs is that the latter provide the capability to manipulate the computed equations though the former only provide a numerical (Fortran, C, Pascal, ...) simulation code including

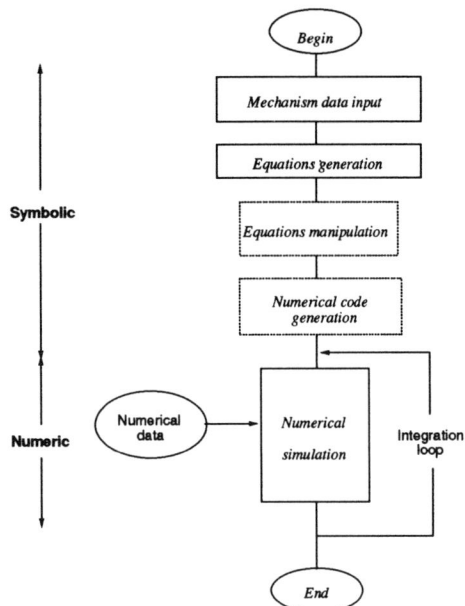

Figure 1: Organization of a symbolic software

subroutines representing the equations.

This architecture presents undoubtedly two great advantages compared to a full numerical one with regard to simulation. First, once the simulation code has been established, the user can perform, by changing the numerical values of the meaningful physical parameters as well as by giving new initial conditions, a wide set of simulations without the waste of time to recompute, after each change, the complete set of equations. Also, symbolic methods can dramatically improve simulation performances which is clearly a very desirable property.

2.1 numerical simulation performances of symbolic softwares

Computational efficiency is often advanced as the primary reason for choosing symbolic multibody programs over their numerical counterparts. The three most important concepts leading to efficient simulation routines are, according to our experiments :

- propagating known numerical values (especially zeros and ones) throughout all invoed expressions,

- unrolling loops to avoid branching and subroutine calls,

- sharing common sub-expressions.

It is well known that lots of mechanical matrices and vectors contain a large number of zeros and ones. For example, any classical cosine matrix contains at least 4 zeros and one one (when Ox, Oy or Oz are typical rotational axis). Proper propagation of these values by themselves can refrain the simulation program from making hundreds of unnecessary operations.

Unneeded subroutine calls represent a severe waste of time during simulations. Let us consider, for example, the multiplication of two cosine matrices $\mathcal{R}_1$ and $\mathcal{R}_2$:

$$\mathcal{R}_1 = \begin{pmatrix} cos(q_1) & -sin(q_1) & 0 \\ sin(q_1) & cos(q_1) & 0 \\ 0 & 0 & 1 \end{pmatrix},$$

$$\mathcal{R}_2 = \begin{pmatrix} 1 & 0 & 0 \\ 0 & cos(q_2) & -sin(q_2) \\ 0 & sin(q_2) & cos(q_2) \end{pmatrix}$$

the first and easiest solution to compute this product consists in calling a *multiply* subroutine

$$call \ multiply(R_1, R_2, 3, 3, 3, Result)$$

which requires one subroutine call and 27 multiplications each time the product has to be computed. This computation can occur up to hundreds of times during a single integration step. Another solution consists in doing the multiplication at the symbolic level (before numerical code generation) and generate the minimum corresponding code to compute the result. Of course, the code generation time will increase but we think that is a small disadvantage compared to the simulation time improvements. Indeed, the result will be computed by a set of 9 assignments requiring no *call* and only 4 multiplications.

$$\begin{aligned} Result_{1,1} &= cos(q_1) \\ Result_{1,2} &= -sin(q_1)cos(q_2) \\ Result_{1,3} &= sin(q_1)sin(q_2) \\ \cdots &= \cdots \\ Result_{3,3} &= cos(q_2) \end{aligned}$$

Last but not least, common sub-expressions sharing may be performed. The detection and storage into intermediate variables of any expression appearing more than once in the whole simulation will save a significant amount of runtime. For example, the previous matrices multiplication will then be translated as follows

$$
\begin{aligned}
f1 &= dsin(q1) \\
f2 &= dcos(q1) \\
f3 &= dcos(q2) \\
f4 &= f2 * f3 \\
f5 &= dsin(q2) \\
f6 &= f1 * f5 \\
f7 &= -f1 * f3 \\
f8 &= -f2 * f5 \\
Result_{1,1} &= f2 \\
Result_{1,2} &= f7 \\
Result_{1,3} &= f6 \\
\cdots &= \cdots \\
Result_{3,3} &= f3
\end{aligned}
$$

Savings may seem limited but one has to remember that the complete set of equations will be computed hundreds or thousands of times (depending of the accuracy required and simulation time interval) during any simulation.

Some compiling experts may argue that good compilers can do the job. That is partially true. In fact, avoiding subroutines calls may be performed by compilers through replacement of each *call* by the corresponding subroutine code but the generated code will still present inefficient loops (a typical matrix multiplication routine involves two loops). For common subexpressions elimination, compilers will only detect *local* redundancy. That is, an expression *expr* appearing two times close one to another will be stored into a register and immediately reused but the limited number of registers will imply that registers will soon be overwritten and a new instance of *expr* encountered a few lines forward will be recomputed.

3 The choice of Maple

Most of the previously mentioned symbolic programs use a specialized and limited symbolic toolbox to derive the equations. Their authors usually say that general computer algebra systems are efficient, powerful, secure, of great interest, ...but they do not use them ! The two principal reasons advanced are, first, that general algebra systems are too large (requiring then a great amount of storage and computational resources) and secondary that the number of useful functions would represent at most one percent of the total amount of functions provided by systems like Maple ([7]), Macsyma ([8]) or Mathematica ([9]).

In our opinion, we think that these reasons are not valid or, at least, no more valid. New algebra systems like Maple even if their library is rather huge, do not require an enormous amount of memory to load their kernel (half a megabyte is usually enough). Systems like Maple can be run on personal computers like IBM-PC or Macintosh as well as on work stations or big systems. Furthermore, during the last years, computing power and calculation speed have fantastically increased while the prices of computers, memory and disks have gone down in the same proportion. These changes in the computer market have disqualified hardware resources limitations. Let's mention, indeed, that our prototype has been successfully tested on a 80386 IBM-PC computer having one megabyte of memory.

Though it is true to say that only a limited number of the existing functions of an algebra system are necessary to derive mechanical equations, we think that other facilities provided by systems like Maple may always be of great interest for a further treatment of the computed equations.

We have chosen to use Maple among other systems for several reasons. First, it is obvious that Maple is one of actual two leading algebra systems on the emerging market of symbolic computation. Then, we are old Maple users and are industrial partner C.N.E.S. also uses it. The last but certainly the most important reason is that Maple is able to handle very large expressions using a reasonable amount of storage space while keeping good runtime performances.

4 Overview of our prototype

Our prototype is now fully operational and widely validated. We called it $\sum$SYGMMAE for *SYmbolical Generation and Manipulation of Mechanical Algebraic Equations*. It is completely written in Maple programming language and works on top of Maple so that any symbolic calculation can be performed during a session. In this section, we shall first expose some of the most important features of the software and then detail its use throughout a simple example in section 5.

4.1 the considered mechanisms

According to its actual state $\sum$SYGMMAE can derive the motion equations of any mechanism composed of an arbitrary number of rigid bodies having a simple kinematic chain structure (no closed loop) and arbitrary distributed prismatic or revolute joints. The first body of the chain may be declared as free in the space and therefore have six degrees of freedom. The system can be subject to any kind of external forces or torques including gravity.

The mechanism may be defined either interactively during a Maple session or by means of a description file containing specific declarative commands and optional Maple assignments to fix, if needed, some of the numerous parameters of the mechanism.

4.2 the formalism employed

During the past decade, the leading and increasingly popular method to derive motion equations is to use Kane's formulation (sometimes called "Lagrange's form of d'Alembert's principle") to obtain the governing dynamical equations of a multibody system. Kane's equations (see [10], [11] or [12] for technical details) have been shown to lead directly to the simplest possible form of the equations. In fact, Kane's equations possess the advantages of both Lagrange's and Newton-Euler's methods avoiding their corresponding disadvantages. Let's notice that Kane's final dynamical equations for a n degrees of freedom system may be written as follows

$$Mass\,\dot{u} + f = 0 \tag{1}$$

Mass being an $n \times n$ matrix, $\dot{u}$ the vector of Kane's *generalized speeds* time derivatives and f a vector including all inertial and external forces.

4.3 equations generation and handling

Once the structural data of the system (number and names of different bodies, type of the links, external forces and torques ...) have been described, $\sum$SYGMMAE can establish the corresponding motion equations using a set of formal variables (see figure 2) attached to each body. The equations are established according to a pure literal matrix form using several specific matrices operators. In spite of Maple existing linear algebra facilities, we have rewritten new operators so that trivial and less trivial simplifications are automatically performed and expressions are derived according to a pseudo canonical form allowing therefore simplifications such as $A * Id = A$, $A * 0 = 0$, $A * 2B = 2(A * B)$, $A.B - B.A = 0$ (scalar product), $A \times B + B \times A = 0$ (cross product) and so on ....

Motion equations can be manipulated in many different ways using our own specialized tools (expansion, transformation according to a set of equations, optimization, linearization, explicit evaluation, numerical code translation ...) as well as with Maple standard library functions.

4.4 numerical code generation

In most cases, the final goal of any mechanical engineer consists in simulating the behavior of the considered mechanism. Our prototype provides the capability to construct complete and ready to compile Fortran or C simulation code. To get a complete simulator code, one successively has to translate the equations into the desired language and generate all the complementary routines to finally integrate the second order differential equation set.

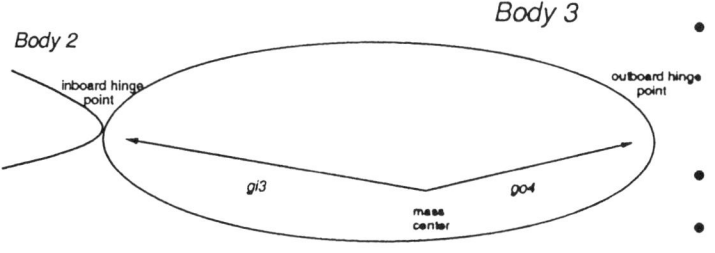

m3   mass of body 3
q3   generalized coordinate of body 3
u3   generalized speed of body 3
l3   vector parallel to joining axis between body 3 and body 2
gi3   vector from mass center of body 3 to body 2 hinge point
go4   vector from mass center of body 3 to body 4 hinge point
in3   central inertia dyadic of body 3

pas3 rotation matrix between body 3 and body 2
fo3   external force upon body 3 center of mass
to3   external torque upon body 3

Figure 2: Formal variables associated to one body

4.4.1 translation into Fortran or C

For Fortran or C equations translation, we use
approximately the same algorithm consisting in
applying the different optimization methods described in section 2.1. This algorithm does not
straightforwardly generate code but constructs a
Macrofort or *MacroC* sequence of instructions .
Macrofort package has been developed by Claude
Gomez ([13]) and I have written *MacroC* ([14])
which is the C language counterpart for *Macrofort*. *Macrofort* is already available in Maple of-
ficial share library and *MacroC* will soon follows.
Without giving technical details on these packages, let us mention that they both allow Fortran or C code generation within Maple including
data structures definitions and declarations, conditional and iterative instructions, subroutines definitions and calls, C preprocessor facilities and so
on . . .

4.4.2 complementary numerical routines

To be complete and ready to compile and use, the
simulation code must include at least four complementary routines :

- a data input routine to assign unknowns
  (bodies' mass, inertia, vectors' components,
  initial conditions, simulation time interval,
  . . .),

- a results output routine,

- a numerical matrix inversion routine to transform system (1) into a form suitable to the
  integration routine that is :

$$\dot{u} = -Mass^{-1}f \qquad (2)$$

Let us mention that such an inversion cannot be performed symbolically because the
Mass matrix coefficients are really too enormous and the matrix size is equal to the
mechanism's number of degrees of freedom
(usually from six to twelve).

- a numerical integration routine to solve the
  final second order differential equation system.

Data input is performed via a special file whose
"mask" is generated by the software. Therefore,
the user only has to fill blank spaces to get a
complete and ready to use data file. Simulation's
results are written into another file according to
a special format which can easily be modified.
For Fortran inversion and resolution, we have chosen to use the well known secure and optimized
routines of the NAG Fortran Library [15] even
when we have written our own C inversion and
resolution routines (see [16]). Fortran integration
routine may be chosen among three : a variable
order, variable step Adams method (default); a
Runge et Kutta - Merson method and a Back-
ward differentiation method for stiff systems.

Let's add, to be complete, that a Unix shell
script is available to perform successively the whole
operations beginning with compilation and ending with a suitable treatment of the results.

5 An example of the use of our prototype

In this section, we give a complete session describing the use of our prototype and its features.
Mechanical expressions are so large that it is not

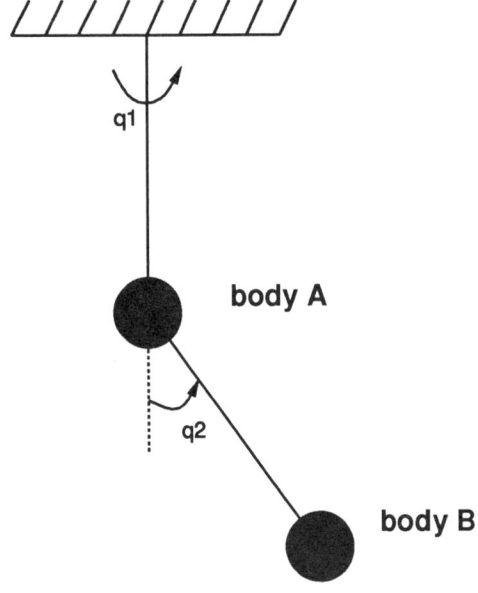

Figure 3: A double pendulum system

possible, in the context of an article, to study
a mechanical system involving more than two or
three degrees of freedom. The considered system
is a double pendulum (see figure 3 as described
by Dan E. Rosenthal in [17]).

```
> # declare a body A linked with the
> # reference inertial body (ground)
> # by a revolute joint
> declare(body, A, rotation);

> # Body B linked to A by a rotation
> declare(body, B, rotation);

> # Equations computation
> sys_diff();

> # Mass matrix and force vectors
> # are stored in arrays called masse
> # and force
> get_dim(masse), get_dim(force);
                [2, 2], [2, 1]

> # First element of mass matrix
> masse[1,1];
&+(&*(&t frame(l1, A, 0),
      frame(inA, A, 0),
      frame(l1, A, 0)),
```

```
  mA &s &*(&t %1, &~ frame(l1, A, 0),
           frame(giA, A, 0)),

  &*(&t frame(l1, A, 0),
     frame(inB, B, 0),
     frame(l1, A, 0)),

  mB &s ((&t &+((-1) &s %1,
              &*((&~ frame(l1, A, 0)),
                 frame(goB, A, 0)),

        (-1) &s ((&~ frame(l1, A, 0)) &*
                 frame(giB, B, 0)))) &*
              &+((-1) &s %1,

        (&~ frame(l1, A, 0)) &*
        frame(goB, A, 0),

        (-1) &s ((&~ frame(l1, A, 0)) &*
                 frame(giB, B, 0))))
)

%1 :=  (&~ frame(l1, A, 0)) &*
       frame(giA, A, 0)

> # frame operators hide some
> # frame changes
> _frame[giB, B, 0];
                &*(pasA, pasB, giB)

> # Let's apply a substitution saying :
> # 1) rotational axis of body A is
> #     parallel to first pendulum axis,
> # 2) inertia matrices are null
> # 3) center of mass of body A is
> #     situated at hinge point with body B
> mat_subs(&x(l1, giA)=0, inA=0,
          inB=0, goB=0, masse[1,1]);

  mB &s &*(&t ((&~ frame(l1, A, 0)) &*
              frame(giB, B, 0)),
           &~ frame(l1, A, 0),
           frame(giB, B, 0))

> # Expand the result
> m11_expand := expand(");
  m11_expand :=
      (-mB) &s &*(&t giB, &t pasB,
              &t pasA,
              (&~ (pasA &* l1)) &^ 2,
              pasA, pasB, giB)
```

```
> # Some global assignments :
> # rotation axis
> l1 := array([0,1,0]):
> l2 := array([0,0,1]):

> # pendulum arms
> giA := array([0,-L,0]):
> giB := array([0,-L,0]):

> # inertia matrices
> inA := 0:
> inB := 0:

> # vector from center of mass
> # of body A to hinge point with
> # body B
> goB := 0;

> # Evaluate the expression using
> # the involved matrices explicit
> # coefficients
> evalmat(m11_expand);
      2      2        2        2
mB L  sin(q2)  (cos(q1)  + sin(q1) )

> # Maple standard routines
> # are still available
> simplify(");
          2      2        2
    mB L  - mB L  cos(q2)

> # The final fully evaluated matrices
> masse_ev:=map(simplify, evalmat(masse));

          [     2      2        2       ]
          [ mB L  - mB L  cos(q2)    0  ]
masse_ev:=[                             ]
          [                          2  ]
          [          0          mB L    ]

> force_ev:=map(simplify, evalmat(force));
                    2
force_ev:=[ 2 mB sin(q2) L  u2 cos(q2) u1,
                    2       2
          - mB cos(q2) L  sin(q2) u1  ]

> # Translate the masse_ev matrix into C
> CC(masse_ev);
                masse_ev
```

```
> # MacroC instructions sequence
> _prog
                    2
    [equalC, C1, L ],

    [equalC, C2, mB C1],

    [equalC, C3, cos(q2)],
                    2
    [equalC, C4, C3 ],

    [equalC, C5, - mB C1 C4],

    [equalC, C6, C2 + C5],

                      [ C6   0  ]
    [matrixm, masse_ev, [         ]]
                      [ 0   C2  ]

> # Effective code generation
> genC([""]);

C1 = L*L;
C2 = mB*C1;
C3 = cos(q2);
C4 = C3*C3;
C5 = -mB*C1*C4;
C6 = C2+C5;
masse_ev[0][0] = C6;
masse_ev[0][1] = 0;
masse_ev[1][0] = 0;
masse_ev[1][1] = C2;

> # Generation of a full and
> # complete simulator including files
> # dpendulum.c      source code
> # dpendulum.data   data file (to fill)
> # dpendulum.res    result file

> simulateur('dpendulum', c);
```

6 Limitations and future extensions

6.1 limitations

The greatest limitation of $\sum_{y}$GMMAE is the size of the considered mechanism. When we say size that means the mechanism total number of degrees of freedom. Like every Maple application we are limited by the total number of terms accepted by the system (that is $2^{17}-1$ terms). Any-the total number of terms accepted by the system (that is $2^{17} - 1$ terms). Anyway, we do not think that this limitation is really severe because motion equations of systems having more than 15 or 20 degrees of freedom are tremendously huge and cannot be handled in any manner. People interested in so large mechanisms will only plan pure numerical simulations and may therefore use another simulation specialized software. Other limitations of the program are linked with the possible extensions described in the next section.

6.2 future extensions

We are developing our prototype and future development axis will be determined with involved users. Nevertheless, we already plan to work in the following directions

- define new types of joints. For the moment, only free, prismatic and revolute joints are allowed. Assuming that any type of joint can be decomposed into prismatic and revolute joints, we plan to extend the joints library,

- write new functions to handle large expressions in good conditions. This may be done by using Maple *optimize* facilities,

- add a new graphic interface for system architecture definition. This X11 interface is already being developed and we wish to use it in the next months,

- model hinge point actuators. This is undoubtedly the most important and interesting point. Actuators can be modeled as external forces or torques acting on the mechanism but this approach cannot be used for spatial mechanism for which actuators' mass and inertia cannot be neglected towards mass and inertia of the involved bodies,

- extend the family of treated bodies to be able to handle flexible bodies. Flexible structures modeling is a hard task and will imply a strong cooperation with good specialists.

7 Conclusion

Advantages of symbolic simulation softwares versus fully numeric ones have been presented. It has been shown that symbolism could substantially increase simulations performances. Furthermore, using a general computer algebra system such as Maple leads to a successful attempt to derive and handle explicit equations of motion for multibody mechanisms. A brief description of our Maple prototype has been given and a limited commented session of its use provided. We are still going on with the development of the prototype and we hope to be able to handle both hinge actuators and flexible structures. We think we have proven that building such a simulation software on top of Maple could lead to an efficient and easy to use program.

Acknowledgments

The work reported in this paper has been partially supported by the Spatial mathematics department of the French Spatial Agency C.N.E.S.

References

[1] R.R. Ryan. Adams multibody system analys software - an introduction. Technical report, Mechanical Dynamics Incorporation, 1989.

[2] Thomas R. Kane and David A. Levinson. *Dynamics : Theory and Applications*. New-York McGraw-Hill Book Company, 1985.

[3] Pierre Coste. *Dynamica user's guide*, matra espace - toulouse edition, 1990.

[4] J. Wittenburg. The dynamics of systems of rigid bodies. 1987.

[5] Symbolic Dyn-Inc. *SD-EXACT/FAST User's manual.* Symbolic Dynamics Incorporation, symbolic dynamics incorporation edition, 1985.

[6] Christophe Garnier and Pascal Rideau. *Manuel d'utilisation du logiciel James (ou Gemmes)*, aerospatiale - cannes edition, 1989.

[7] Bruce W. Char, Keith O. Geddes, Gaston H. Gonnet, Benton L. Leong, Michael B. Monagan, and Stephen M. Watt. *Maple V Language Reference Manual.* Springer-verlag, 1992.

[8] Symbolics-Inc. *Macsyma User's Guide.* Symbolics-Inc., 11 Cambridge Center, Cambridge MA 02142, 1988.

[9] S. Wolfram. *Mathematica — A System for Doing Mathematics by Computer.* Addison-Wesley Publishing Company, Redwood City, California, 1988.

[10] Duc Minh Tran. Une présentation de la méthode de Kane pour la formulation des équations du mouvement. *La recherche Aérospatiale*, 3, 1991.

[11] Thomas R. Kane and David A. Levinson. The use of Kane's dynamical equations in robotics. *The International Journal of Robotics Research*, 2(3), 1983.

[12] Thomas R. Kane, P. W. Likins, and David A. Levinson. *Spacecrafts Dynamics.* Mc Graw-Hill New-York, 1983.

[13] Claude Gomez. Macrofort : a fortran code generator in maple. Rapport technique 119, Institut National de Recherche en Informatique et en Automatique, 1990.

[14] Patrick Capolsini. MacroC : C code generation within Maple. Rapport technique 151, Institut National de Recherche en Informatique et en Automatique - Laboratoire Informatique Signaux et Systèmes de l'Université de Nice Sophia-Antipolis, 1992.

[15] NAG Incorporation. *The NAG Fortran librairy Manual*, 1990.

[16] W.H. Press, B.P. Flannery, S.A. Teukolsky, and W.T. Vetterling. *Numerical recipes in C : the art of scientific computing.* Cambridge University Press, 1988.

[17] Dan E. Rosenthal. An order n formulation for robotic systems. Technical report, Symbolic Dynamics, Mountain View, California, 1987.

biography

Patrick Capolsini is a Ph.D. student in the *Safir* project which is common to INRIA, CNRS and University of Nice.

He is interested in symbolic computation of mechanical equations and their optimized numerical transcription. He belongs to the *I3S laboratory*, 250 Avenue Albert Einstein - 06560 VAL-BONNE - FRANCE (capolsin@safir.unice.fr).

SENSITIVITY ANALYSIS OF NONLINEAR PHYSICAL SYSTEMS USING MAPLE

Stephen Carr, Gordon J. Savage
Dept. of Systems Design Engineering, University of Waterloo, Waterloo ON, Canada

Abstract

Sensitivities are the partial derivatives of system responses with respect to the parameters of a design model. These have long been used as measures of the "importance" of individual parameters in the model, but are also required in reliability and quality analyses, and for design optimization. Sensitivities are typically calculated numerically, either estimated using finite differences, or calculated directly using techniques derived from network theory, such as the well-known Adjoint method. Direct methods make use of the true derivatives of system equations and have the advantage of requiring many fewer system analyses than the finite difference methods; they are also exact for linear system models. Symbolic computation software, such as Maple, can further increase the computational efficiency of the direct sensitivity analysis methods. The gain arises from the ability to solve the sensitivity models in algebraic form. The algebraic solution need only be performed *once*. The numerical work is thus reduced from repeated generation of Jacobians and inverses at each operating point to simple evaluation of the algebraic sensitivity expressions. For nonlinear system models numerical solution, *e.g.* by the Newton-Raphson method, cannot be avoided but many of the advantages of using Maple remain. Sensitivity analysis using Maple is demonstrated for a simple, nonlinear pipe network system.

Mathematical Computation with Maple V:
Ideas and Applications
Tom Lee, Editor
©1993 Birkhäuser Boston

1 Introduction

Sensitivity analysis of engineering models is an essential part of the design process. The 'sensitivity factors', or simply 'sensitivities', are the partial derivatives of the responses of the system model with respect to its parameters, or 'inputs'. Sensitivities have long been used to measure the "importance" of individual parameters in the model [7, 8], but are also required in reliability and quality analyses [2, 17, 12], and design optimization [10, 16].

Sensitivities are typically calculated numerically, either estimated using finite differences, or calculated directly using techniques derived from systems theory [1] , such as the well-known Adjoint method. Direct methods make use of the true derivatives of system equations and have the advantage of requiring many fewer system analyses than the finite difference methods; they are also exact for linear system models. In this paper we demonstrate a direct method, Graph-Theoretic Sensitivity Models (GTSMs)[14], using Maple.

The mathematical model of an engineering system is often in the form of a system of nonlinear equations

$$\mathbf{h}(\mathbf{v}, \mathbf{z}) = \mathbf{0} \qquad (1)$$

where $\mathbf{z}$ and $\mathbf{v}$ are vectors of system responses and inputs, respectively. A system model in this form is implicit in the response variables, $\mathbf{z}$, and the solution must be obtained iteratively by, for example, the Newton-Raphson method.

Traditionally, the sensitivity problem is addressed *after* the system equations (1) have been

---

[1] Also known as network theory, or graph-theoretic modelling.

assembled and much of the correspondence between the mathematics and the system structure is lost. Under these circumstances it is not obvious how the sensitivity equations are to be constructed. However, in graph-theoretic sensitivity models, derivatives are taken at the component level first, then the system-level sensitivity models are assembled. Using the standard methods of systems theory [1, 4, 11, 15, 13], the 'nominal' system equations, 'First-Order Sensitivity (FOS)' equations, 'Second-Order Sensitivity (SOS)' equations, etc. , can all be built up automatically from the system description. Each successively higher-order sensitivity model incorporates the solutions of the preceding models. The GTSM method allows sensitivity analysis to be viewed as a simple matrix algebra problem, without appealing to the abstract notion of an "adjoint" system [14]. Also, because it is a direct method, calculation of the complete sensitivity information of a chosen order is obtained from a single solution of each of the models in the hierarchy.

Symbolic computation has been found to be very useful for the construction and solution of both nominal and sensitivity system models. This paper attempts to demonstrate this utility using Maple [6] software. A brief introduction to some systems theory concepts and the solution of nonlinear system models is given in Section 2. (For details, interested readers are directed to the cited references.) Section 3 describes the development of the system sensitivity equations. In Section 4 sensitivity analysis methods are applied to a simple nonlinear pipe network. Finally, we provide some concluding remarks.

For convenience in dealing with matrix expressions, we invoke operations from matrix calculus and the 'Kronecker product' operator, "$\otimes$". Relevant definitions are given in Appendix A. As notation, we denote n^{th} order derivatives of a matrix $\mathbf{A}$ with respect to another matrix $\mathbf{B}$ with the superscript notation $\mathbf{A}^{(nB)}$. By convention, we write all vectors as *column* vectors but denote the nth derivative of a vector by a vector as $\partial\mathbf{a}/\partial\mathbf{b}^T \equiv \mathbf{a}^{(nb)}$. The derivative is written with respect to the *transpose* of $\mathbf{b}$ in order to correspond with the familiar definition of the Jacobian

matrix when $n = 1$.

2 Nominal System Model

2.1 A Little Systems Theory

The 'system variables' which describe an engineering system are its characterizing measurements. These variables fall into two generalized categories, 'through' and 'across'. Examples of through variables, denoted as $\mathbf{y}$, are current, flow rate, force, etc. . Across variables, denoted $\mathbf{x}$, include voltage, displacement, temperature, and pressure drop as examples.

According to systems theory, a system is completely described by three sets of equations. Two sets of equations, known collectively as the 'interconnection equations', describe the constraints imposed on the system variables by the interconnection of the components in a system. The interconnection equations are the 'circuit equations' and the 'vertex equations', which are generalizations of Kirchhoff's Voltage Law and Kirchhoff's Current Law, respectively. The interconnection equations are always linear, and may be written in matrix form in terms of the 'incidence matrix', $\mathbf{A}$, a matrix of ± 1's and 0's, which describes the system topology. The system topology is represented by a 'system graph' of directed line segments which indicate the polarity of through and across measurements.

The vertex equations are

$$\mathbf{A}\mathbf{y} = \mathbf{0} \qquad (2)$$

and, a special form of the circuit equations is the 'nodal transformation equations'

$$\mathbf{x} = \mathbf{A}^T\mathbf{x}_n \qquad (3)$$

which introduces an additional set of across variables, $\mathbf{x}_n$, the 'nodal' variables. The nodal variables represent measurements between the components' terminals of interconnection and an external datum, e.g. gauge pressure or displacement with respect to an inertial frame of reference.

The remaining set of equations needed to describe the system is the governing equations, or 'terminal equations', of the system's components.

119

The terminal equations of a component are derived by characterizing measurements taken with respect to the terminals of the component. A simple electrical resistor, for example, has the single terminal equation, $v = r\,i$ (or $x = r\,y$). Any nonlinearities in the system model arise from the terminal equations. For example, the flow in a pipe as a function of the pressure drop is given by the empirical Hazen-Williams equation:

$$q = g\,h^{0.54} \qquad (4)$$

where q is the pipe flow (through variable) and h is the pressure drop (across variable). The coefficient, g is a function of other input parameters:

$$g = \frac{Cd^{2.63}}{kl^{0.54}} \qquad (5)$$

with C, the Hazen-Williams pipe roughness; d, the pipe diameter; l, the pipe length; k, a constant determined by the units used. A component which has a single terminal equation, as in the above examples, has only two terminals with which to connect to other components. In general we can write the terminal equations of two-terminal 'constitutive' components as a system of equations

$$\mathbf{y}_c = \mathbf{G}\,\mathbf{f}(\mathbf{x}_c) \qquad (6)$$

where the subscript c indicates system variables associated with constitutive components; $\mathbf{f}(\mathbf{x}_c)$ is a vector of nonlinear functions of $\mathbf{x}_c$. The matrix $\mathbf{G}$ in equation (6) is a diagonal matrix with elements $g_i(\mathbf{u})$, where vector $\mathbf{u}$ is a set of inputs. The more general case of multi-terminal components is discussed in [13].

The terminal equations for the through and across 'excitations', or 'drivers', provide specified functions of time, and they are respectively

$$\mathbf{y}_t = \mathbf{y}_{t0}(t)\,; \quad \mathbf{x}_a = \mathbf{x}_{a0}(t) \qquad (7)$$

where subscripts t, a, and 0 denote "through", "across", and "specified". Excitations often arise from uncontrolled environmental conditions acting on the system, but may also be components which supply energy to the system, *e.g.* voltage sources or pumps.

Together, the two sets of interconnection equations plus the set of terminal equations of all components constitute a determinate system where the number of equations equals the number of unknown through and across variables. The combined system of equations is known as the 'system equations', or simply the 'system model'. Various formulations of the system model are possible, varying from the complete solution for all unknown through and across variables to more compact formulations which directly provide only a subset of the unknown system variables. Often, we are just interested in the nodal variables. A convenient and compact formulation is the 'mixed nodal tableau (MNT)'(also known as the 'modified nodal' formulation) where all unknown system variables have been eliminated except the nodal variables and the through variables corresponding to the across excitations.

2.2 Solution of Nonlinear Systems: the Newton-Raphson System

The solution of the nominal system is the 'nominal response', which is determined for any set of fixed input values. If the terminal equations are nonlinear, the nominal response must be determined iteratively by *e.g.* the Newton-Raphson Method.

The general nonlinear system model can be written in matrix form as in equation (1). We construct the MNT formulation from the basic equations (2), (3), (6) and (7) as

$$\mathbf{h} = \left[\begin{array}{c} \mathbf{A}_c\mathbf{G}\mathbf{f}(\mathbf{A}_c^T\mathbf{x}_n) + \mathbf{A}_a\mathbf{y}_a + \mathbf{A}_t\mathbf{y}_{t0} \\ \mathbf{A}_a^T\,\mathbf{x}_n - \mathbf{x}_{a0} \end{array} \right] = \left[\begin{array}{c} 0 \\ 0 \end{array} \right] \qquad (8)$$

where partitions of the incidence matrix, $\mathbf{A}$, corresponding to the various component types are indicated by subscripts. The first row in (8) is just the vertex equations written in partitioned form, with terminal equations and nodal transformation equations (where possible) substituted into the first and third terms. The second row is the terminal equations of the across excitations, with the nodal transformation substituted for $\mathbf{x}_a$. The response variables $\mathbf{z} = [\mathbf{x}_n\ \mathbf{y}_a]^T$ are implicit in (8).

To solve (1) or (8) using the Newton-Raphson method, we expand $\mathbf{h}$ in a linear Taylor series about operating point z^k as

$$\mathbf{h} \approx \mathbf{h}(\mathbf{z}^k) + h^{(1z)}(\mathbf{z}^k)(\mathbf{z} - \mathbf{z}^k) = \mathbf{0} \qquad (9)$$

Equation (9) may be written as a linear system of equations

$$\mathbf{L}_{NR}(\mathbf{z}^k)\,\Delta\mathbf{z}^k = \mathbf{r}_{NR}(\mathbf{z}^k) \qquad (10)$$

where $\mathbf{L}_{NR}$ is the Jacobian, $h^{(1z)}(\mathbf{z}^k)$; $\mathbf{r}_{NR}(\mathbf{z}^k)$ is $-\mathbf{h}(\mathbf{z}^k)$; and $\Delta\mathbf{z}^k \equiv \mathbf{z} - \mathbf{z}^k$. The Newton-Raphson system, (10), may be solved for the Newton-Raphson "step", $\Delta\mathbf{z}^k$, which suggests the iteration scheme $\mathbf{z}^{k+1} = \mathbf{z}^k + \Delta\mathbf{z}^k$.

However, it is more convenient to use the techniques of systems theory to build up the Newton-Raphson system directly from component information [4, 5]. If we rearrange (10) to put only $\mathbf{z}$ on the LHS, we get

$$\mathbf{L}_{NR}(\mathbf{z}^k)\,\mathbf{z} = \mathbf{r}_{NR}(\mathbf{z}^k) + \mathbf{L}_{NR}(\mathbf{z}^k)\,\mathbf{z}^k \qquad (11)$$

The response, $\mathbf{z}$, of the nominal system is obtained at the convergence of the iteration

$$\mathbf{z}^{k+1} = \mathbf{L}_{NR}^{-1}(\mathbf{z}^k)\left\{\mathbf{r}_{NR}(\mathbf{z}^k) + \mathbf{L}_{NR}(\mathbf{z}^k)\,\mathbf{z}^k\right\} \quad (12)$$

Writing (8) in the form of (11) we obtain the Newton-Raphson MNT formulation

$$\begin{bmatrix} \mathbf{A}_c\,\mathbf{G}\,\mathbf{f}^{(1x_c)}(\mathbf{A}_c^T\,\mathbf{x}_n)\,\mathbf{A}_c^T & \mathbf{A}_a \\ \mathbf{A}_a^T & 0 \end{bmatrix} \begin{bmatrix} \mathbf{x}_n \\ \mathbf{y}_a \end{bmatrix} =$$

$$\begin{bmatrix} -\mathbf{A}_c\left\{\mathbf{G}\,\mathbf{f}(\mathbf{A}_c^T\,\mathbf{x}_n^k) - \mathbf{G}\,\mathbf{f}^{(1x_c)}(\mathbf{A}_c^T\,\mathbf{x}_n^k)\,\mathbf{A}_c^T\,\mathbf{x}_n^k\right\} \\ \mathbf{x}_{a0} \\ -\mathbf{A}_t\,\mathbf{y}_{t0} \end{bmatrix}_{(2\times1)}$$

$$(13)$$

The Newton-Raphson MNT (13) may be built using systems-theoretic concepts by developing modified terminal equations for the constitutive components. For the nonlinear constitutive equations of the form given in equation (6), and treating $\mathbf{G}$ as constant, we obtain the first-order Taylor expansion

$$\begin{aligned} \mathbf{y}_c &= \left[\mathbf{G}\,\mathbf{f}^{(1x_c)}(\mathbf{x}_c^k)\right]\mathbf{x}_c & (14) \\ &+ \left\{\mathbf{G}\,\mathbf{f}(\mathbf{x}_c^k) - \left[\mathbf{G}\,\mathbf{f}^{(1x_c)}(\mathbf{x}_c^k)\right]\mathbf{x}_c^k\right\} \end{aligned}$$

where $\mathbf{G}$ is treated as constant, and the derivative of the matrix product (6) is defined in Appendix A; operating point values $\mathbf{x}_c^k$ may be obtained from $\mathbf{z}^k$ by the nodal transformation (3). The term in braces in equation (14) may be considered to be a "pseudo" through excitation which appears in the Newton-Raphson system model but not in the real (physical) system. Note also that some care must be taken with the signs of terms in (14) so that the y_c's change sign only when x_c's change sign [5].

The nominal interconnection equations, (2) and (3), and the terminal equations of excitations (7), remain unchanged in the formulation of the Newton-Raphson MNT, equation (13).

3 Sensitivity System Models

In sensitivity analysis, we are interested in the sensitivities of responses, $\mathbf{z}$, with respect to the set of the inputs of the system model. Inputs may be physical or geometric parameters of system components or values of system excitations. We write the inputs as a vector of length m as

$$\mathbf{u}^T = [u_1,\, u_2, \ldots, u_m]$$

The 'design variables' of the system constitute some subset of the inputs, e.g. the geometric parameters.

The nature of the nominal system model outlined in the previous section can be extended to models that have derivatives of the through and across variables with respect to the input vector, $\mathbf{u}$. A separate 'sensitivity model' is developed for each order of derivatives required.

3.1 FOS System Equations

Let us now take the derivatives of the vertex and nodal transformation equations, (2) and (3), in the nominal model. We proceed by using the rule for the derivative of a matrix product given in Appendix A. Since the incidence matrix, $\mathbf{A}$, is constant, for the vertex equations we obtain

$$
\mathbf{A} \begin{bmatrix} \partial y_1/\partial u_1 & \partial y_1/\partial u_2 & \dots & \partial y_1/\partial u_m \\ \partial y_2/\partial u_1 & \partial y_2/\partial u_2 & \dots & \partial y_2/\partial u_m \\ \vdots & & \ddots & \\ \partial y_p/\partial u_1 & \partial y_p/\partial u_2 & \dots & \partial y_p/\partial u_m \end{bmatrix}
$$
$$
= \begin{bmatrix} 0 & 0 & \dots & 0 \\ 0 & 0 & \dots & 0 \\ \vdots & & \ddots & \\ 0 & 0 & \dots & 0 \end{bmatrix}
$$

In the superscript notation described above we can simply write

$$\mathbf{A}\mathbf{y}^{(1u)} = \mathbf{0} \qquad (15)$$

Similarly, for the nodal transformation equations we obtain

$$\mathbf{x}^{(1u)} = \mathbf{A}^T \mathbf{x}_n^{(1u)} \qquad (16)$$

Note that we now have m *sets* of vertex equations and m *sets* of nodal transformation equations written in compact form; the *vectors* of system variables in the nominal interconnection equations now become *matrices* of FOS system variables.

Now we derive the FOS terminal equations for constitutive components of the form given in equation (6). We define the vector $\mathbf{v} = [\mathbf{x}_{a0}, \mathbf{g}(\mathbf{d}), \mathbf{y}_{t0}]^T$ where the elements of vector $\mathbf{g}(\mathbf{d})$ are the diagonal elements of matrix $\mathbf{G}$ in (6) which are functions of physical properties and design variables, $\mathbf{d}$; $\mathbf{x}_{a0}$ are across excitations and $\mathbf{y}_{t0}$ are through excitations. The vector $\mathbf{u}$ is defined as $\mathbf{u} = [\mathbf{x}_{a0}, \mathbf{d}, \mathbf{y}_{t0}]^T$. We define the transformation between $\mathbf{v}$ and $\mathbf{u}$ as $\mathbf{v} = \mathbf{v}(\mathbf{u})$. The set of inputs $\mathbf{v}$ contain variables such as $\mathbf{g}$ which we use only for notational convenience. We require sensitivities with respect to the actual measurable quantities contained in $\mathbf{u}$. Applying the Chain and Product rules of matrix calculus (see Appendix A) to equation (6) we obtain

$$\mathbf{y}_c^{(1u)} = \left[\mathbf{G}\,\mathbf{f}^{(1x_c)}\right]\mathbf{x}_c^{(1u)} + \mathbf{G}^{(1v)}\left(\mathbf{v}^{(1u)} \otimes \mathbf{f}\right) \quad (17)$$

Equation (17) is *linear* in the FOS constitutive across variables, $\mathbf{x}_c^{(1u)}$, since the second term (a pseudo through excitation) may be evaluated once

the solution to the nominal system is available. Note that (17) is the same as (14) except for the form of the pseudo through excitation.

FOS terminal equations for the excitations are much simpler. Applying the Chain Rule to equations (7),

$$\mathbf{y}_t^{(1u)} = \mathbf{y}_{t0}^{(1v)}\mathbf{v}^{(1u)} \; ; \quad \mathbf{x}_a^{(1u)} = \mathbf{x}_{a0}^{(1v)}\mathbf{v}^{(1u)} \quad (18)$$

Given the FOS interconnection equations, (15) and (16), and the FOS terminal equations, (17) and (18), the FOS system model can be formulated (using *e.g.* the Mixed Nodal Tableau) as a *linear* system of equations in the FOS response variables:

$$\mathbf{L}(\mathbf{u}, \mathbf{z})\,\mathbf{z}^{(1u)} = \mathbf{R}_{FOS}(\mathbf{u}, \mathbf{z}) \qquad (19)$$

where $\mathbf{L}$ is the same as $\mathbf{L}_{NR}$ in (11) and only the RHS is different. The FOS response variables, $\mathbf{z}^{(1u)}$, form a matrix with m columns, where m is the length of vector $\mathbf{u}$.

3.2 SOS System Equations

We obtain the SOS system equations by differentiating the FOS equations again with respect to $\mathbf{u}$.

The SOS vertex and nodal transformation equations simply become

$$\mathbf{A}\mathbf{y}^{(2u)} = \mathbf{0} \; ; \quad \mathbf{x}^{(2u)} = \mathbf{A}^T \mathbf{x}_n^{(2u)} \qquad (20)$$

The SOS terminal equations for the constitutive components become, after some manipulation,

$$
\begin{aligned}
\mathbf{y}_c^{(2u)} = {} & \left(\mathbf{G}\mathbf{f}^{(1x_c)}\right)\mathbf{x}_c^{(2u)} && (21)\\
& + \Big\{ \mathbf{G}\mathbf{f}^{(2x_c)}\big(\mathbf{x}_c^{(1u)} \otimes \mathbf{x}_c^{(1u)}\big) \\
& + \mathbf{G}^{(2v)}\big(\mathbf{v}^{(1u)} \otimes \mathbf{v}^{(1u)} \otimes \mathbf{f}\big) \\
& + \mathbf{G}^{(1v)}\Big[\big(\mathbf{v}^{(2u)} \otimes \mathbf{f}\big) \\
& + \mathbf{U}\big(\mathbf{f}^{(1x_c)}\mathbf{x}_c^{(1u)} \otimes \mathbf{v}^{(1u)}\big) \\
& + \big(\mathbf{v}^{(1u)} \otimes \mathbf{f}^{(1x_c)}\mathbf{x}_c^{(1u)}\big)\Big] \Big\}
\end{aligned}
$$

All the terms encompassed by the braces in (21) may be evaluated given the nominal and FOS solutions and constitute a pseudo through excitation. Note that (21) is also the same as (14) and

122

(17) except for the form of the pseudo through excitation. The second-order derivatives of the excitations usually vanish.

The SOS system model may also be formulated as a linear system of equations of the form

$$\mathbf{L}(\mathbf{u}, \mathbf{z})\, \mathbf{z}^{(2u)} = \mathbf{R}_{SOS}(\mathbf{u}, \mathbf{z}) \qquad (22)$$

where, again, $\mathbf{L} = \mathbf{L}_{NR}$. The SOS response variables, $\mathbf{z}^{(2u)}$, form a matrix with m^2 columns.

4 Pipe Network Example

As an example, consider the simple pipe network system shown in Figure 1 a). The through

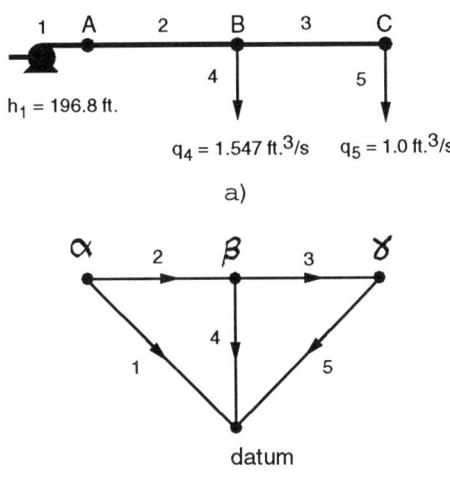

Figure 1: a) Simple Pipe Network
b) System Graph

and across variables for this system are $\mathbf{y} = [q_1, q_2, q_3, q_4, q_5]^T$ and $\mathbf{x} = [h_1, h_2, h_3, h_4, h_5]^T$, *i.e.* the flows through and heads across the components shown in Figure 1 a). Orientations of the through and across measurements are represented by the system graph shown in Figure 1 b). If the Mixed Nodal Tableau is used to formulate the system equations, the solution variables are $\mathbf{z} = [h_\alpha, h_\beta, h_\gamma, q_1]^T$, where the heads are nodal

across measurements taken at nodes A through C, and q_1 is the flow through the fixed-head pump.

The system excitations are the demand flows at nodes B and C, and the head supplied by the pump at node A. These are all assumed to be constant: $q_4 = 1.547(ft.^3/s)$; $q_5 = 1.0(ft.^3/s)$; $h_1 = 196.8(ft.)$.

The constitutive equation for the pipes is the Hazen-Williams equation, given in (4) and (5), where $C = 100$, $d = 10(in.)$, $k = 852000$, and $l = 3280(ft.)$. C and d are parameters, while k and l are considered to be fixed constants in the design. In order to take advantage of the notational convenience of (5), we define two sets of input variables: $\mathbf{v} = [h_1, g_2, g_3, q_4, q_5]^T$ and $\mathbf{u} = [h_1, C_2, C_3, d_2, d_3, q_4, q_5]^T$. The transformation from $\mathbf{u}$ to $\mathbf{v}$ is given by (5) for g_2 and g_3, and the remaining variables are unchanged in the transformation. Written in the matrix form of (6), the constitutive terminal equations are

$$\begin{bmatrix} q_2 \\ q_3 \end{bmatrix} = \begin{bmatrix} g_2 & 0 \\ 0 & g_3 \end{bmatrix} \begin{bmatrix} h_2 \,|h_2|^{-\frac{23}{50}} \\ h_3 \,|h_3|^{-\frac{23}{50}} \end{bmatrix} \qquad (23)$$

where $h_i^{0.54}$ in (4) has been rewritten as $|h_i|^{0.54} h_i/|h_i| = h_i |h_i|^{-0.46}$ in order to ensure that the signs (directions) of q_2 and q_3 change with those of their respective pressure drops [5].

4.1 Nominal Solution

Since the constitutive terminal equations are nonlinear, we build the Newton-Raphson system (11) and solve for $\mathbf{z}$ iteratively using (12).

The constitutive terminal equations, (14), are

$$\mathbf{y}_c = \begin{bmatrix} \dfrac{27}{50} \dfrac{g_2}{|h_1 - h_\beta^k|^{\frac{23}{50}}} & 0 \\ 0 & \dfrac{27}{50} \dfrac{g_3}{|h_\beta^k - h_\gamma^k|^{\frac{23}{50}}} \end{bmatrix} \mathbf{x}_c$$

$$+ \begin{bmatrix} \dfrac{23}{50} \dfrac{g_2(h_1 - h_\beta^k)}{|h_1 - h_\beta^k|^{\frac{23}{50}}} \\ \dfrac{23}{50} \dfrac{g_3(h_\beta^k - h_\gamma^k)}{|h_\beta^k - h_\gamma^k|^{\frac{23}{50}}} \end{bmatrix} \qquad (24)$$

where operating point values $\mathbf{h}_2$ and $\mathbf{h}_3$ have been replaced with nodal variables via the nodal transformation (3), and h_α has been replaced by the specified across driver value, h_1. The pseudo through excitation is the second term in (24). The Newton-Raphson MNT, equation (13), may be built in symbolic form using Maple. The solution of the Newton-Raphson MNT provides the iteration (12)

$$
\begin{aligned}
h_a^{k+1} &= h_1 \\
h_\beta^{k+1} &= \frac{1}{27g_2}\left(-50q_4|h_1 - h_\beta^k|^{\frac{23}{50}}\right. \\
&\quad + 50g_2h_1 - 23g_2h_\beta^k \\
&\quad \left. - 50q_5|h_1 - h_\beta^k|^{\frac{23}{50}}\right) \\
h_\gamma^{k+1} &= \frac{1}{27g_2g_3}\left(-50g_3q_4|h_1 - h_\beta^k|^{\frac{23}{50}}\right. \\
&\quad + 50g_2g_3h_1 - 50g_2q_5|h_\beta^k - h_\gamma^k|^{\frac{23}{50}} \\
&\quad \left. - 23g_2g_3h_\gamma^k - 50g_3q_5|h_1 - h_\beta^k|^{\frac{23}{50}}\right) \\
q_1^{k+1} &= -q_4 - q_5
\end{aligned}
\tag{25}
$$

(shown here as separate equations).

For starting guess $h_\alpha = 196.8$, $h_\beta = 195$, $h_\gamma = 190$, and $q_1 = -2.5$, equation (25) stabilizes with errors less than 0.001 after four iterations to give the solution $[196.8, 154.7, 147.3, -2.547]^T$. Note that in this problem h_α is just the across excitation and q_1 is determined by the vertex equations applied at the datum node in Figure 1 b), so the iteration (25) is not a function of these variables. The full solution for all across and through variables may be retrieved, if desired, by back-substitution into the nodal transformation equations (3) and the constitutive terminal equations (23).

4.2   FOS solution

The FOS system model is constructed as described in Section 3. Of particular interest are the FOS terminal equations of the constitutive components which are obtained by differentiating the nominal terminal equations using equation (17). These equations are the same as (24), above, except for the pseudo through excitation term:

$$
\begin{bmatrix}
0 & \dfrac{d_2^{\frac{263}{100}}(h_1-h_\beta)}{(kl)^{\frac{27}{50}}|h_1-h_\beta|^{\frac{23}{50}}} & 0 & & & & \\[2ex]
0 & 0 & \dfrac{d_3^{\frac{263}{100}}(h_\beta-h_\gamma)}{(kl)^{\frac{27}{50}}|h_\beta-h_\gamma|^{\frac{23}{50}}} & & & & \\[2ex]
\dfrac{263}{100}\dfrac{d_2^{\frac{263}{100}}C_2(h_1-h_\beta)}{(kl)^{\frac{27}{50}}|h_1-h_\beta|^{\frac{23}{50}}} & 0 & & & 0 & 0 \\[2ex]
0 & \dfrac{263}{100}\dfrac{d_3^{\frac{263}{100}}C_3(h_\beta-h_\gamma)}{(kl)^{\frac{27}{50}}|h_\beta-h_\gamma|^{\frac{23}{50}}} & & & 0 & 0
\end{bmatrix}
\tag{26}
$$

a (2×7) matrix where the columns contain derivatives with respect to h_1, C_2, C_3, d_2, d_3, q_4, and q_5, respectively.

Since all sensitivity system models are linear, we may obtain the exact symbolic FOS solution as a (4×7) matrix

$$
\mathbf{z}^{(1u)} =
\begin{bmatrix}
1 & 0 & 0 \\[1ex]
1 & \dfrac{50}{27}\dfrac{d_2^{\frac{263}{100}}(h_1-h_\beta)}{g_2(kl)^{\frac{27}{50}}} & 0 \\[2ex]
1 & \dfrac{50}{27}\dfrac{d_2^{\frac{263}{100}}(h_1-h_\beta)}{g_2(kl)^{\frac{27}{50}}} & \dfrac{50}{27}\dfrac{d_3^{\frac{263}{100}}(h_\beta-h_\gamma)}{g_3(kl)^{\frac{27}{50}}} \\[2ex]
0 & 0 & 0
\end{bmatrix}
$$

$$
\begin{matrix}
0 & 0 \\[1ex]
\dfrac{263}{54}\dfrac{d_2^{\frac{263}{100}}C_2(h_1-h_\beta)}{g_2(kl)^{\frac{27}{50}}} & 0 \\[2ex]
\dfrac{263}{54}\dfrac{d_2^{\frac{263}{100}}C_2(h_1-h_\beta)}{g_2(kl)^{\frac{27}{50}}} & \dfrac{263}{54}\dfrac{d_3^{\frac{263}{100}}C_3(h_\beta-h_\gamma)}{g_3(kl)^{\frac{27}{50}}} \\[2ex]
0 & 0
\end{matrix}
$$

$$
\begin{matrix}
0 & 0 \\[1ex]
-\dfrac{50}{27}\dfrac{|h_1-h_\beta|^{\frac{23}{50}}}{g_2} & -\dfrac{50}{27}\dfrac{|h_1-h_\beta|^{\frac{23}{50}}}{g_2} \\[2ex]
-\dfrac{50}{27}\dfrac{|h_1-h_\beta|^{\frac{23}{50}}}{g_2} & -\dfrac{50}{27}\dfrac{g_2|h_\beta-h_\gamma|^{\frac{23}{50}}+g_3|h_1-h_\beta|^{\frac{23}{50}}}{g_2g_3} \\[2ex]
-1 & -1
\end{matrix}
\tag{27}
$$

where rows correspond to h_α, h_β, h_γ, and q_1, and columns contain their partial derivatives with respect to h_1, C_2, C_3, d_2, d_3, q_4, and q_5, respectively. This expression may be evaluated once the nominal solution is available, since it requires h_β and h_γ. Errors in h_β and h_γ are carried over from the nominal solution.

The complete FOS solution for all through and across variables in the system model may be recovered by substitution of (27) into (16) and (26).

4.3 SOS Solution

The SOS solution may also be obtained in exact symbolic form. For this example, the SOS MNT solution, $z^{(2u)}$, is a matrix of size (4×49), and is too large to show here. The full SOS solution may also be recovered by substitution into the SOS nodal transformation equations and the SOS constitutive equations.

Of particular interest for optimization of the design are the second-order partial derivatives of the heads at nodes B and C with respect to the pipe diameters. The non-zero second partial derivatives are

$$\frac{\partial^2 h_\beta}{\partial d_2^2} = \frac{\partial^2 h_\gamma}{\partial d_2^2} =$$
$$\frac{-263C_2(h_1-h_\beta)\left(20251C_2 d_2^{\frac{163}{50}}(kl)^{\frac{27}{50}} - 4401 d_2^{\frac{63}{100}} g_2(kl)^{\frac{27}{25}}\right)}{145800 g_2^2(kl)^{\frac{81}{50}}}$$

$$\frac{\partial^2 h_\gamma}{\partial d_3^2} =$$
$$\frac{-263C_3(h_\beta-h_\gamma)\left(20251C_3 d_3^{\frac{163}{50}}(kl)^{\frac{27}{50}} + 4401 d_3^{\frac{63}{100}} g_3(kl)^{\frac{27}{25}}\right)}{145800 g_3^2(kl)^{\frac{81}{50}}}$$

5 Conclusions

Maple has been shown to be useful for the construction and solution of sensitivity models in the following respects:

1. Maple's symbolic differentiation capabilities are used to differentiate the component's constitutive equations to produce the Newton-Raphson model of the nominal nonlinear system and the sensitivity models.

2. Maple programs can be written to build up the nominal and sensitivity models in symbolic form using systems theory (given a data file describing the types of components in the system and their interconnections).

3. Solution of the nominal system model may be obtained exactly in symbolic form for linear systems. For nonlinear systems, it is possible to obtain the symbolic expression for for the Newton-Raphson iteration, equation (12).

4. Since the sensitivity models of any order are always *linear*, exact symbolic solutions can always be obtained, even though symbolic solutions are not generally available for nonlinear nominal system models. The availability of symbolic solutions facilitates study of the behavior of the sensitivities as functions of the inputs.

5. Given the symbolic sensitivity solutions, numeric values may be obtained by simple substitution of the nominal responses and the input values. For nonlinear physical systems, the calculated nominal response values are inexact.

A Matrix Calculus Definitions

The following definitions have been adapted from Graham [9]. These operations are easily implemented in Maple procedures [3]. In the following, let $\mathbf{A}$ be $p \times q$ matrix and $\mathbf{B}$ be $r \times s$. Let $\mathbf{v}$ be a column vector of length n and $\mathbf{u}$ be of length m.

A.1 Derivative of a Matrix With Respect to a Vector

$$\mathbf{A}^{(1v)} = \frac{\partial \mathbf{A}}{\partial \mathbf{v}^T} = \left[\begin{array}{cccc} \frac{\partial \mathbf{A}}{\partial v_1} & \frac{\partial \mathbf{A}}{\partial v_2} & \cdots & \frac{\partial \mathbf{A}}{\partial v_n} \end{array}\right]_{p \times nq} \quad (28)$$

Note that in this definition the vector $\mathbf{v}$ is transposed. This produces a matrix which has the same form as the Jacobian matrix of a vector function.

Similarly for higher-order derivatives,

$$\mathbf{A}^{(2v)} = \left[\begin{array}{cccc} \frac{\partial \mathbf{A}^{(1v)}}{\partial v_1} & \frac{\partial \mathbf{A}^{(1v)}}{\partial v_2} & \cdots & \frac{\partial \mathbf{A}^{(1v)}}{\partial v_n} \end{array}\right]_{p \times n^2 q} \quad (29)$$

A.2 Derivative of a Matrix With Respect to a Matrix

$$\mathbf{A}^{(1B)} = \frac{\partial \mathbf{A}}{\partial \mathbf{B}} = \left[\begin{array}{cccc} \frac{\partial \mathbf{A}}{\partial b_{11}} & \frac{\partial \mathbf{A}}{\partial b_{12}} & \cdots & \frac{\partial \mathbf{A}}{\partial b_{1s}} \\ \vdots & & \ddots & \\ \frac{\partial \mathbf{A}}{\partial b_{r1}} & \frac{\partial \mathbf{A}}{\partial b_{r2}} & \cdots & \frac{\partial \mathbf{A}}{\partial b_{rs}} \end{array}\right]_{pr \times qs} \quad (30)$$

A.3 Product Rule

When $q = r$ and for differentiation with respect to a vector, $\mathbf{v}$, we can define

$$(\mathbf{AB})^{(1v)} = \mathbf{A}^{(1v)}(\mathbf{I}_n \otimes \mathbf{B}) + \mathbf{AB}^{(1v)} \qquad (31)$$

where $\mathbf{I}$ is an $n \times n$ identity matrix.

A.4 Chain Rule

Let elements of $\mathbf{A}$ be functions of $\mathbf{v}$ and let elements of $\mathbf{v}$ be functions of $\mathbf{u}$. Then

$$\mathbf{A}^{(1u)} = \mathbf{A}^{(1v)}(\mathbf{v}^{(1u)} \otimes \mathbf{I}_q) \qquad (32)$$

Applying the Chain Rule to a matrix product, for $q = r$

$$(\mathbf{AB})^{(1u)} = \mathbf{A}^{(1v)}(\mathbf{v}^{(1u)} \otimes \mathbf{B}) + \mathbf{AB}^{(1u)} \qquad (33)$$

A.5 Kronecker Matrix Product

$$\mathbf{A} \otimes \mathbf{B} = \begin{bmatrix} a_{11}\mathbf{B} & a_{12}\mathbf{B} & \dots & a_{1q}\mathbf{B} \\ a_{21}\mathbf{B} & a_{22}\mathbf{B} & \dots & a_{2q}\mathbf{B} \\ \vdots & & \ddots & \\ a_{q1}\mathbf{B} & a_{q2}\mathbf{B} & \dots & a_{qq}\mathbf{B} \end{bmatrix}_{pr \times qs}$$

$$(34)$$

A.6 Kronecker Product Rule

The derivative of a Kronecker product with respect to a vector , $\mathbf{v}$, is

$$(\mathbf{A} \otimes \mathbf{B})^{(1v)} = \mathbf{A}^{(1v)} \otimes \mathbf{B} + \mathbf{U}(\mathbf{B}^{(1v)} \otimes \mathbf{A}) \qquad (35)$$

where $\mathbf{U}$ is a permutation matrix defined as

$$\mathbf{U} = \sum_{i=1}^{p}\sum_{j=1}^{r} \mathbf{E}_{ij} \otimes \mathbf{E}_{ij}^{T} \qquad (36)$$

and $\mathbf{E}_{ij}$ is an 'elementary' matrix of size $p \times r$ with single non-zero element $e_{ij} = 1$.

References

[1] G.C. Andrews and H.K. Kesavan. The vector-network model: a new approach to vector dynamics. *Journal of Mechanisms and Machine Theory*, 10:57–75, 1975.

[2] S.M. Carr and G.J. Savage. Comparison of chance-constrained and structural reliability formulations of reliability-based design. Submitted to: Reliability Engineering and System Safety, May 1993.

[3] S.M. Carr and G.J. Savage. Maple procedures for matrix calculus. Internal Report, Department of Systems Design, University of Waterloo, May 1993.

[4] M. Chandrashekar and H.K. Kesavan. Graph-theoretic models for pipe network analysis. *Journal of the Hydraulics Division, ASCE*, 98(HY2):345–364, 1972.

[5] M. Chandrashekar and K.H. Stewart. Sparsity oriented analysis of large pipe networks. *Journal of the Hydraulics Division, ASCE*, 101(HY4):341–355, 1975.

[6] B.W. Char, K.O. Geddes, G.H. Gonnet, B.L. Leong, M.B. Monagon, and S.M. Watt. *Maple V Language Reference Manual*. Springer-Verlag, New York, 1992.

[7] A. Deif. *Sensitivity Analysis in Linear Systems*. Springer-Verlag, Berlin, 1986.

[8] M. Eslami and R.S. Marleau. Theory of sensitivity of network: A tutorial. *IEEE Transactions on Education*, 32(3):319–334, 1989.

[9] A. Graham. *Kronecker Products and Matrix Calculus with Applications*. Ellis-Howard Ltd., Chichester, 1981.

[10] E.J. Haug, K.K. Choi, and V. Komkov. *Design Sensitivity Analysis of Structural Systems*, volume 177 of *Mathematics in Science and Engineering*. Academic Press, New York, 1986.

[11] N.C. Lind. Analysis of structures by system theory. *Journal of the Structural Division, ASCE*, 88(ST2):1–22, 1962.

[12] H.O. Madsen, S. Krenk, and N.C. Lind. *Methods of Structural Safety*. Prentice-Hall, Englewood Cliffs, NJ, 1986.

[13] P.H.O'N. Roe. *Networks and Systems.* Addison-Wesley, Reading, Mass., 1966.

[14] G.J. Savage. Automatic formulation of higher-order sensitivity models. To appear: Civil Engineering Systems, 1993.

[15] G.J. Savage and H.K. Kesavan. The graph-theoretic field model - I: modelling and formulations. *Journal of the Franklin Institute,* 307(2):107–147, 1979.

[16] J.N. Siddall. *Optimal Engineering Design.* Marcel Dekker, Inc., New York, 1982.

[17] R. Spence and R.S. Soin. *Tolerance Design of Electronic Circuits.* Addison-Wesley, New York, 1988.

Biographies

Gordon Savage is an Associate Professor, and Stephen Carr is a doctoral student, in the Department of Systems Design Engineering at the University of Waterloo. Research interests of both authors include modelling and simulation of engineering systems using graph theory, with applications to design based on reliability and quality criteria.

The authors may be reached at

Department of Systems Design
University of Waterloo
Waterloo, Ontario
Canada, N2L 3G1
gjsavage@watserv1.uwaterloo.ca
scarr@watserv1.uwaterloo.ca

EXACT CALCULATION OF THE KAPLAN-MEIER BIAS USING MAPLE SOFTWARE

Brenda Gillespie, Justine Uro
Department of Biostatistics, University of Michigan, Ann Arbor MI, USA

1. Introduction

In many medical studies and industrial experiments, the outcome of interest is the time to some event such as death, disease recurrence or component failure. The investigators often want to find the distribution of these times, as well as characteristics of the distribution such as the mean, median, dispersion and skewness. In such studies it is usually impractical to wait until all items or subjects have failed, and those which have not failed at the time of data analysis are right-censored, meaning that the times recorded are lower bounds on the actual failure times.

In the censored data setting, the observed data are t_i, the observed failure or censoring time, and δ_i, the censor code which equals 1 if t_i is a failure time and zero if it is a censoring time. A data set is of the form $(t_1, \delta_1), (t_2, \delta_2), \ldots, (t_n, \delta_n)$. For the purposes of this paper, we assume no tied values in the sample data. To simplify notation, let the sample t_i's be ordered such that $t_1 < t_2 < \ldots < t_n$.

The distribution of these "survival times" is usually expressed in terms of the survivor function, $S(t) = P(T > t)$, which gives the probability of surviving longer than time t. Here, T is the random variable representing the time to failure. If data were all uncensored, then this function is estimated by the proportion of items which have not failed by time t, i.e., the empirical survivor function. When the data include censored values, then a modified estimator is required, and the Kaplan-

Meier (KM) estimator is commonly employed. The KM estimator is a nonparametric maximum-likelihood estimator of the survivor function in the presence of censored data, and reduces to the empirical survivor function when censoring is not present. The Kaplan-Meier estimator of $S(t)$ is given by:

$$\hat{S}(t) = \prod_{i:t_i \leq t} \left[\frac{n-i}{n-i+1} \right]^{\delta_i}, \qquad (1)$$

where $i = 1, \ldots, n$. Note that $\hat{S}(t)$ is a right-continuous, non-increasing step function starting at $\hat{S}(t) = 1$ for t less than the first uncensored point, and with jumps down at the uncensored points. If the largest data value is uncensored, then $\hat{S}(t)$ jumps down to zero at that point. Each term of the KM product is the estimated conditional probability of surviving to time t_i, given survival to time t_{i-1}. If these were the actual probabilities, their product would yield the unconditional probability of surviving to time t. The KM estimator can be intuitively understood from this perspective.

The bias of an estimator is the expected value (i.e., mean of the estimator under repeated sampling) minus the true value being estimated. Since an entire survivor function is being estimated, the bias of the estimator must be evaluated at each point of time, t. The KM estimator is known to be biased, but the bias as a function of t cannot be expressed in closed form unless strong assumptions are made about the relationship between the survivor and censoring distributions (Chen, Hollander and Langberg, 1982). Previous investigations of the bias of the KM estimator in the general case have used approximations based

Mathematical Computation with Maple V:
Ideas and Applications
Tom Lee, Editor

on computer simulations. However, it is possible to express the bias exactly in terms of multiple integrals. If these integrals are explicitly evaluable, then one can compute the bias exactly using symbolic mathematical software such as Maple. We have developed a Maple program to perform these calculations. Higher moments of the KM estimator (yielding variance, skewness, etc.) may be computed in a similar manner using this program.

The program is based on expressing the probability of each possible KM estimate value as a multiple integral, given known survivor and censoring distributions. We enumerate all possible cases, i.e., all possible values that the KM estimate could assume. For a given sample size n and a fixed point in time t, the value of the KM estimator depends on how many sample values are less than t, and the pattern of censored and uncensored values among those observations. If there are $r \leq n$ sample values less than t, there will be 2^r situations to evaluate since each value may be either exact or censored. To enumerate these cases, we generate a sequence of binary integers from zero to $2^r - 1$ (left-filled with zeros), where in each binary number, a zero digit represents a censored observation and a one represents an exact observation. Each binary string defines a multiple integral, with integrand components specified by the binary sequence. Evaluation of the multiple integral yields the probability of the associated KM estimate, from which the KM expectation and the bias can be calculated.

This application of Maple has resulted in an exact result where only an approximate result was previously available. Sample size is a limiting problem since the total number of integrals to be evaluated is $2^{n+1} - 2$, where n is the sample size. Some improvements to the algorithm which reduce the number of calculations are currently under investigation.

Statistical details of the calculation are given in Section 2. A description of the Maple implementation is given in Section 3. Some efficiency considerations are discussed in Section 4, and conclusions are given in Section 5.

2. Calculation of the Kaplan-Meier bias

In the original definition of the KM estimator (Kaplan and Meier, 1958), the estimator is not defined past the last data point if that point is censored. To compute the bias, however, some definition must be given in this situation. So-called completion methods have been developed by Efron (1967), Gill (1980), Brown, Hollander and Korwar (1974) and Moeschberger and Klein (1985). Estimators based on completion methods are bounded below by the method of Efron, who suggested $\hat{S}(t) = 0$ for $t > t_n$, the last observation, whether or not t_n is censored; they are bounded above by the estimator of Gill, who suggested $\hat{S}(t) = \hat{S}(t_n)$ for $t > t_n$. We denote the Gill version by $\hat{S}_G(t)$, and note that this is the definition that was used in expression (1). We illustrate the calculation of the KM bias with the Gill completion method in this paper, although modification of the bias calculation for other completion methods is straightforward. In the sequel, we use the notation $\hat{S}_G(t)$ whenever the statement applies only to the KM estimator with the Gill completion method. If the statement applies with any completion method, we use the notation $\hat{S}(t)$.

The bias of $\hat{S}(t)$ as a function of t, $B(t)$, is formally defined as $B(t) = E[\hat{S}(t)] - S(t)$, where $E[\cdot]$ is the expectation (mean) operator. The bias of $\hat{S}(t)$ has been discussed by Efron (1967), Meier (1975), Chen et al. (1982), Geurts (1985, 1987), Klein (1988), and Gillespie et al. (1992). The bias of $\hat{S}_G(t)$ is positive (Klein, 1988) and increases to an asymptote as $t \to \infty$.

Let X and Y denote the (unobserved) random variables for the failure and censoring times, respectively, and assume that X and Y are independent. The observed random variables are $T = \min(X, Y)$ and δ, the censor code, which equals 1 if $X \leq Y$ and zero otherwise. Sample values of the given random variables are denoted (t_i, δ_i). The method for computing moments of the KM estimator in the general case can be given in terms of integrals of the joint distribution of T and δ. The number of multiple integrals which must be calculated equals the number of possible values that $\hat{S}(t)$ may take at any time point with various sequences of failures and censored values. For example, to calculate the KM bias at some fixed time t, first suppose that t falls between the ith and $(i+1)$th observation, i.e., $t_i < t < t_{i+1}$. Then there are 2^i possible unique sequences of 0s and 1s prior to t, and thus 2^i potentially unique values of $\hat{S}(t)$. Taking the sum over the $i = 0 \ldots n$ possible positions for t relative to the t_is and summing the geometric series, we have

$$\sum_{j=0}^{n} 2^j = \frac{1 - 2^{n+1}}{1 - 2} = 2^{n+1} - 1.$$

This is the number of cases which must be considered, but the single case when all data values are greater than t does not require integral evaluation. Thus, the number of multiple integrals which must be calculated is $2^{n+1} - 2$, which will be computationally feasible for small n. Table 1 below gives some values of $2^{n+1} - 2$ for illustration.

Table 1. Number of integrals evaluated for several values of n

n	Number of Integrals
2	6
5	62
10	2,046
20	2,097,150

The cumulative distribution function (c.d.f.) of a random variable, X, is the probability that X is larger than some fixed value, say x, as a function of x, and is usually given an upper case letter, e.g., $F(x)$. Note that the c.d.f. is 1 minus the survivor function for continuous random variables. The density function is the derivative of the c.d.f.; the probability that X falls into a certain interval can be obtained by integrating the density function over that interval. With these informal definitions, denote the c.d.f.s of X and Y by $F(x)$ and $G(y)$, and the densities by

$f(x)$ and $g(y)$, respectively. Let the support (i.e., the range of values where the random variable has positive density) of $f(x)$ and $g(y)$ be $(0, u_x)$ and $(0, u_y)$, respectively, where u_x and u_y could be infinity. Then the joint c.d.f. of T and δ (the joint c.d.f. of a continuous and a discrete random variable) is given by:

$$F_{T,\delta}(t, \delta = 0) = \int_0^t \int_y^{u_x} f(x)\, g(y)\, dx\, dy,$$

$$F_{T,\delta}(t, \delta = 1) = \int_0^t \int_x^{u_y} f(x)\, g(y)\, dy\, dx$$

with the densities, $f_{T,\delta}(t, \delta = 0)$ and $f_{T,\delta}(t, \delta = 1)$ obtained by differentiation with respect to t. The regions of integration above are illustrated in Figure 1 below, where region **A** applies to $F_{T,\delta}(t, \delta = 0)$ and region **B** applies to $F_{T,\delta}(t, \delta = 1)$.

Figure 1.   Regions of integration

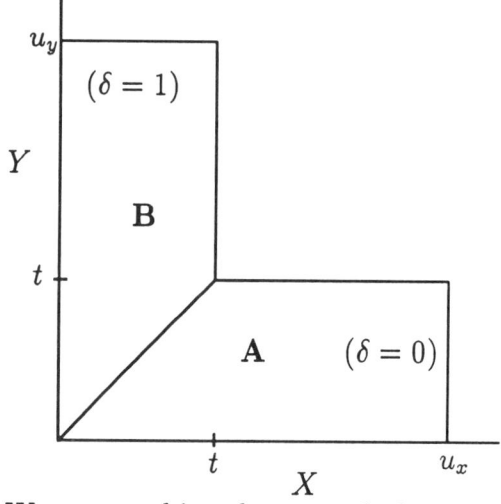

We can combine the two sub-densities into a single expression for the joint distribution of T and δ using indicator functions. For compact notation we define $f_a(t) = f_{T,\delta}(t, \delta = 0)$ and $f_b(t) = f_{T,\delta}(t, \delta = 1)$.

$$f_{T,\delta}(T, \delta) = f_a(t)\, I[\delta = 0] + f_b(t)\, I[\delta = 1]$$

where $I[\cdot]$ is an indicator function which equals one if the condition is true and zero otherwise. This joint distribution will be used below.

The method of obtaining the expected value of $\hat{S}(t)$ uses conditioning: Let the set of possible sequences of 0s and 1s prior to t be denoted by Ω, where Ω has $2^{n+1}-1$ elements, generically denoted ω. Then

$$E[\hat{S}(t)] = E_\Omega[E[\hat{S}(t)|\omega]]$$

$$= \sum_{i=1}^{2^{n+1}-1} E[\hat{S}(t)|\omega_i] \cdot P(\omega_i). \qquad (2)$$

Given ω_i, $E[\hat{S}(t)|\omega_i]$ is simply $\hat{S}(t)|\omega_i$, which is a KM calculation following (1). Calculating $P(\omega_i)$, however, requires evaluation of a multiple integral.

The method is first illustrated for $n = 1$ and $n = 2$, and then generalized to larger n. Note that the time, t, for which we are calculating the KM bias is fixed, and T_i, the data values, are the random variables. For $n = 1$, $\Omega = \{\phi, \{0\}, \text{and } \{1\}\}$, where ϕ denotes the case where $t < T_1$ (i.e., t is prior to the first and only data value), and the other two cases refer to $T_1 \leq t$ (i.e., the data value, T_1, is $\leq t$, and δ_1 is either zero or one). Denoting the above elements of Ω as ω_1, ω_2, and ω_3, respectively, and substituting into expression (1) to compute $\hat{S}(t)|\omega_i$, the resulting values are 1, 1, and 0, respectively. Thus,

$$E[\hat{S}_G(t)] = 1 \cdot P(\omega_1) + 1 \cdot P(\omega_2) + 0 \cdot P(\omega_3)$$

where $P(\omega_1) = P(t < T_1)$

$$= P(X > t) \cdot P(Y > t)$$

$$= (1 - F(t))(1 - G(t))$$

and $P(\omega_2) = P(t \geq T_1, \delta = 0)$

$$= \int_0^t f_{T,\delta}(t, \delta = 0)\, dt$$

$$= F_{T,\delta}(t, \delta = 0).$$

Similarly, $P(\omega_3) = F_{T,\delta}(t, \delta = 1)$, although it need not be calculated since its coefficient is zero. Note that calculating $P(\omega_1)$, i.e., the probability that all data values are greater than t, does not require integration. For example, suppose X and Y are both exponentially distributed random variables with parameter $\lambda = 1$, so that $F(x) = 1 - e^{-x}$, $G(y) = 1 - e^{-y}$, $f(x) = e^{-x}$, $g(y) = e^{-y}$, $P(\omega_1) = e^{-t}e^{-t} =$

e^{-2t}, and $P(\omega_2) = \int_0^t \int_y^\infty e^{-x}e^{-y}\, dx\, dy = 1 - \frac{1}{2}e^{-2t}$. Thus,

$$E[S_G(t)] = e^{-2t} + (1 - \frac{1}{2}e^{-2t}) = 1 + \frac{1}{2}e^{-2t}.$$

Since $S(t) = 1 - F(t) = e^{-t}$, the bias as a function of t is then obtained as $B_G(t) = (1 + \frac{1}{2}e^{-2t}) - e^{-t}$.

For $n = 2$, Ω has 7 elements: ϕ, $\{0\}$, $\{1\}$, $\{0,0\}$, $\{0,1\}$, $\{1,0\}$, and $\{1,1\}$. The first element corresponds to the case $t < T_1 < T_2$, the next two elements to the case $T_1 \leq t < T_2$, and the last four elements to the case $T_1 < T_2 \leq t$. The corresponding values of $E[\hat{S}(t)|\omega_i]$ (which is simply $\hat{S}(t)|\omega_i$) are given in Table 2 below.

Table 2: Values of $\hat{S}_G(t)|\omega_i$ for $n = 2$

ω	$\hat{S}_G(t)|\omega$
ϕ	1
$\{0\}$	1
$\{1\}$	0.5
$\{0,0\}$	1
$\{0,1\}$	0
$\{1,0\}$	0.5
$\{1,1\}$	0

For $n = 1$ we needed the joint distribution of T and δ. For $n = 2$ we need the joint distribution of T_1, T_2, δ_1 and δ_2, where δ_1 and δ_2 are the censor codes corresponding to T_1 and T_2. Again, this is a partly discrete, partly continuous joint distribution with a separate term for each unique sequence of 0s and 1s. For $n = 2$ there are four terms, each of which is a continuous function of t_1 and t_2. (Recall that T_1 and T_2 refer to the ordered values, i.e., the order statistics.) The joint distribution of $(T_1, \delta_1, T_2, \delta_2)$ is given by the following product:

$$n!(f_a(t_1)I[\delta_1 = 0] + f_b(t_1)I[\delta_1 = 1])$$

$$\cdot (f_a(t_2)I[\delta_2 = 0] + f_b(t_2)I[\delta_2 = 1]).$$

This expression is of the form $n!(a_1 + b_1)(a_2 + b_2)$; multiplying the terms together to get $n!(a_1 a_2 + a_1 b_2 + b_1 a_2 + b_1 b_2)$, we have the sub-densities for every arrangement of censored and exact values (i.e., (0,0),(0,1),(1,0),(1,1)). Then, for example, if $\omega_i = \{1\}$, which indicates that $T_1 \le t < T_2$ and $\delta_1 = 1$, then $P(\omega_i) = \int_t^\infty \int_0^t f_b(t_1)(f_a(t_2) + f_b(t_2)) \, dt_1 dt_2$. Note that for positions for which $t < T_i$, the terms for f_a and f_b may be summed.

In general, if T_i is the ith order statistic with associated censor code δ_i, the joint distribution of $(T_1, \delta_1, \ldots, T_n, \delta_n)$ is given by $n!$ times the following product:

$$(f_a(t_1) \, I[\delta_1 = 0] + f_b(t_1) \, I[\delta_1 = 1])$$

$$\cdot(f_a(t_2) \, I[\delta_2 = 0] + f_b(t_2) \, I[\delta_2 = 1])$$

$$\cdots$$

$$\cdot(f_a(t_n) \, I[\delta_n = 0] + f_b(t_n) \, I[\delta_n = 1]).$$

Multiplying out, this expression equals a sum of terms, each corresponding to a unique sequence of 0's and 1's.

For example, if $n = 4$ and $\{\delta_1, \ldots, \delta_4\} = \{0, 1, 1, 0\}$, the corresponding term is $4! \, f_a(t_1) \, f_b(t_2) \, f_b(t_3) \, f_a(t_4)$. This term can be integrated to give probabilities of corresponding ωs. To find $P(\omega)$ for cases in which t is not larger than all data values, the terms for f_a and f_b are summed for the positions after t since $\hat{S}(t)$ only depends on censor codes prior to t. For example, to find $P(\omega)$ for $\omega = \{0, 1, 1\}$, with $n = 4$, we have

$$P(\omega) = P[t_3 < t < t_4, \text{ and}$$

$$\delta_1 = 0, \, \delta_2 = 1, \delta_3 = 1, \delta_4 = \{0 \text{ or } 1\}]$$

$$= \int_t^\infty \int_0^t \int_0^{t_3} \int_0^{t_2} 4! \, f_a(t_1) \, f_b(t_2) \, f_b(t_3)$$

$$\cdot(f_a(t_4) + f_b(t_4)) \, dt_1 \, dt_2 \, dt_3 \, dt_4.$$

The Kaplan-Meier estimate, given ω is:

$$\hat{S}(t) \mid \omega = 1 \cdot \frac{2}{3} \cdot \frac{1}{2} = \frac{1}{3}.$$

After calculating all such terms, we can compute

$$E[\hat{S}(t)] = \sum_{\omega \in \Omega} E[\hat{S}(t) \mid \omega] \cdot P(\omega),$$

from which the bias is obtained by subtracting $S(t)$.

3. Maple implementation

The first part of the program defines the failure and censoring densities, $f(x)$ and $g(y)$, and the sub-densities ($f_{T,\delta}(t, \delta)$ for $\delta = 0, 1$), which we refer to as $f_a(t)$ and $f_b(t)$, respectively.

```
pdf1:=proc(x)   f(x)  end:
pdf2:=proc(y)   g(y)  end:
fa:=proc(t)
  diff(int(int(pdf1(x)*pdf2(y),
    x=y..ux),y=0..t),t)
  end:
fb:=proc(t)
  diff(int(int(pdf1(x)*pdf2(y),
    y=x..uy),x=0..t),t)
  end:
fab:=proc(t)
  fa(t)+fb(t)
  end:
```

The structure of the program after these function definitions is illustrated in Figure 2 below.

Computation of the expected value of $\hat{S}(t)$ entails enumerating the elements ω of Ω. This is done by considering two nested do .. od loops, as shown below.

```
for i from 0 by 1 to n do
  for j from 0 by 1 to (2^i)-1 do
    bj:=convert(j,binary):
    lbj:=length(bj):
    if j=0 then lbj:=1 fi:
    x:=1:
    Gshat:=1:

    (other program statements)
  od:
od:
```

The outer loop, indexed by i, is the number of sample values that are less than t. (It is actually more efficient to start the outer loop at i=1, with i=0 handled as a special case.) The inner loop indexed by j generates as a binary string all possible patterns of exact and censored values (1s and 0s) among the i observations that are less than t. The string is generated by converting the integers from 0 to 2^i-1 to a binary string, `bj`. If `j=0`, corresponding to all observations censored prior to t, then the length of `bj` must be set equal to one. If the length of the string generated, `lbj`, is less than `i` then `i-lbj` zero(s) could be added to the left of the string to yield `i` digits. As an alternative to actually left-filling with zeros, we loop over these positions separately, as shown in Figure 2. Thus, left-filling with zeros is implied but not actually performed.

Since the calculations are built as successive products, we initialize the probability, `x`, to 1 and the expected KM (Gill version) estimate, `Gshat`, to 1.

Figure 2. Structure of the program

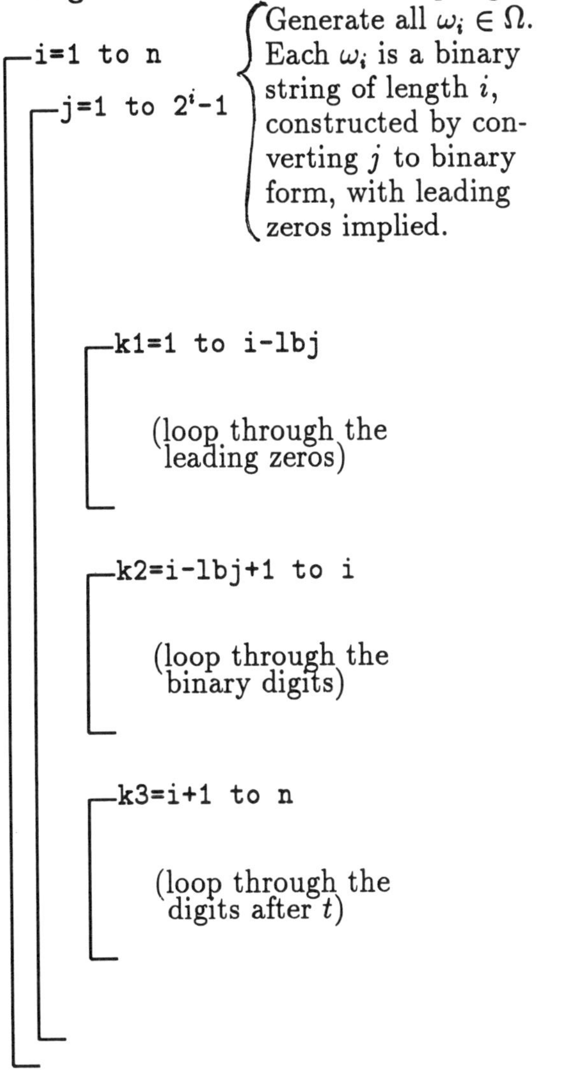

```
                 ⎧ Generate all ωᵢ ∈ Ω.
 ┌i=1 to n       │ Each ωᵢ is a binary
 │               │ string of length i,
 │  ┌j=1 to 2ⁱ-1 ⎨ constructed by con-
 │  │            │ verting j to binary
 │  │            │ form, with leading
 │  │            ⎩ zeros implied.
 │  │
 │  │
 │  │  ┌k1=1 to i-lbj
 │  │  │
 │  │  │
 │  │  │  (loop through the
 │  │  │   leading zeros)
 │  │  └
 │  │
 │  │  ┌k2=i-lbj+1 to i
 │  │  │
 │  │  │
 │  │  │  (loop through the
 │  │  │   binary digits)
 │  │  └
 │  │
 │  │  ┌k3=i+1 to n
 │  │  │
 │  │  │
 │  │  │  (loop through the
 │  │  │   digits after t)
 │  │  └
 │  └
 └
```

For each $\omega \epsilon \Omega$, the KM estimate and the corresponding probability have to be computed. To compute the probability, an n-fold integral is evaluated. The limits of integration for the probability depend on `i`, and the integrand of the multiple integral depends on the pattern of the 0s and 1s and on `i`. Each computation requires individually stepping through all `n` positions to iteratively calculate the KM estimate and to iteratively perform the multiple integration. The `n` steps are performed within `do .. od` loops, split into three cases (see Figure 2). The first case, indexed by `k1`, loops through any leading zeros; the second case, indexed by `k2`, loops through the binary digits from the first 1 until the `i`th digit; the third case, indexed by `k3`, loops through the positions following `i`. In the first two loops, the KM estimator is recursively calculated. In all three loops the calculation of $P(\omega_i)$ is continued.

In the first loop, indexed by `k1`, no update of the KM estimator is needed since censored observations do not change the value of the KM estimate. The contribution of these censored observations to the probability $P(\omega_{ij})$ is obtained by calculating an `(i-lbj)`-fold integral as follows, where the appropriate t variable (t_1, t_2, etc.) is obtained by concatenating t with the looping index, `k1`:

```
for k1 from 1 by 1 to i-lbj do
   x:=int(x*fa(t.k1),
      (t.k1)=0..(t.(k1+1)))
od:
```

The lower limit of integration is always 0 and the upper limit of the `k1`th integral is t_{k1+1}.

Next we loop over the index `k2`, which consists of the `(i-lbj+1)`th digit to the `i`th digit of the binary string constructed. The KM calculation is performed (`Gshat`), which only requires updating at an uncensored value, i.e., a censor code equal to 1. The KM estimate decreases by a factor of $\frac{(n-jn)}{(n-jn+1)}$ if a 1 is encountered and does not change if a zero is encountered. The calculation of $P(\omega_{ij})$, is continued as `lbj` steps of the multiple integration. We define `fs:=fa(t.k2)` when a 0 is encountered and `fs:=fb(t.k2)` when a 1 is encountered. The variable `d` stands for delta

133

(δ), which is the censor code, and d takes the value of each digit of the binary string, one by one. Since the substring operator cannot be applied to an integer (i.e., bj), we concatenate the single letter q to the front of bj. The positions read in the binary string are then from 2 to lbj+1.

```
for k2 from i-lbj+1 by 1 to i do
 d:=substring(q.bj,(k2-i+lbj+1)..
     (k2-i+lbj+1)):
 if d='0' then
  fs:=fa(t.k2):
 else
  fs:=fb(t.k2):
  Gshat:=Gshat*((n-k2)/(n-k2+1)):
 fi:
 if k2<i then
  x:=int(x*fs,(t.k2)=0..
     (t.(k2+1))):
 elif k2=i then
  x:=int(x*fs,(t.i)=0..t):
 fi:
od:
```

Observe that the lower limit of integration is always zero and the upper limit of the k2th integral is t_{k2+1} except for the ith integral which has upper limit t.

We are finally left with the n-i outermost integrals which correspond to those observations which are greater than t (Figure 2). It does not matter whether these observations are exact or censored: The calculation of the KM estimate only depends on observations up to the ith, and for the calculation of $P(\omega_{ij})$, integration is over the sum, fab(t)=fa(t)+fb(t) which is the marginal density of t. The do .. od loop, indexed by k3 gives the contribution of these observations to $P(\omega_{ij})$.

```
for k3 from i+1 by 1 to n  do
 if k3<N then
  x:=int(x*fab(t.k3),
     (t.k3)=t..(t.(k3+1))):
 else
  x:=int(x*fab(t.n),(t.n)=t..ux):
 fi:
od:
```

The lower limit of integration is t while the upper limit is t_{k3+1}, except for the last

integral whose upper limit is the upper bound of the support of the failure distribution.

At this point, we now have the KM estimate (Gill version) and the associated probability, given by $P(\omega_{ij}) = n!x$ for this specific element of Ω. The two are multiplied together and accumulated in the variable Gm (for the Gill version) as given in equation (2).

After the outer loops are completed, the bias is attained as the difference between the true survival time based on the actual failure density and the expected KM estimate (accumulated in Gm).

Some results from this program are presented in Table 3 below. In the situations where exact values can be checked (Case 1 in the table), the Maple results agree exactly with the theoretical results, as expected. In other situations (Case 2 in the table), the Maple results give much more accuracy than was available from published simulations.

Table 3: Results from a Maple program

Case 1: X and Y both have the exponential($\lambda = 1$) distribution

	Exact bias (Maple)	Exact bias (Chen et al.)
$t = 1/2$	-.0017	-.0017
$t = 1$	-.0339	-.0339
$t = 2$	-.0742	-.0742

Case 2: X distributed exponential(1), Y distributed Uniform$(0, b)$, where b is specified to give $P(\text{censoring})=1/3$

	Exact bias (Maple)	Simulated bias (Geurts)
$t = 1/2$	-.000006	-.00
$t = 1$	-.004279	-.00
$t = 2$	-.045279	-.05

Although the current program is only usable with density functions which are repeatedly integrable, it may be possible to use Taylor series or other integrable approximations for densities which are not explicitly integrable.

4. Efficiency considerations

We were able to run this program with n as large as 9. We were able to get results for $n = 10$ by running the outer loop separately for `i=1 to 9` and `i=10`, and summing the results. The program for $n = 9$ took several hours on a SUN Sparcstation 2 with 16 megs of RAM.

It is unlikely that significant savings can be obtained by modifications of the existing program. We have considered switching the order of the two outer loops in order to save some intermediate calculations for future steps. An alternative which may be more promising may be inferred from Meier (1975), who suggested that contributions to the Kaplan-Meier bias only arise from the situation when t is beyond the last data point. If we could limit ourselves to this case, we would save a power of 2 in the number of calculations required. In addition, another power of 2 could be saved since all strings ending in 1 (when $t > T_n$) have KM estimate $\hat{S}(t) = 0$, so $P(\omega_i)$ need not be calculated. Thus, the number of integrals to be calculated would be 2^{i-1} instead of $2^{i+1} - 2$. These improvements would potentially raise the evaluable sample size to between 10 and 15. Since the KM bias becomes negligible for sample sizes larger than, say, 25, there is still a gap between current computing power and desired results.

5. Conclusions

The Maple program described here carries out an exact calculation for a problem which was previously thought to require computer simulation. The Maple methods which proved useful were the ability to generate all possible cases using incremented binary strings, and then using the generated sequences to specify the components of multiple integrals.

Drawbacks of the Maple implementation include the large memory (RAM) requirements, and the length of time required to run the program, both of which limit the possible sample size. In addition, this method only yields exact results when the original functions are explicitly integrable. The advantages of the Maple implementation are greater accuracy and the satisfaction of obtaining an exact rather than an approximate answer.

References

1. Brown, B.W., Hollander, M. and Korwar, R.M. (1974). Nonparametric tests of independence for censored data, with applications to heart transplant studies. In *Reliability and Biometry, Statistical Analysis of Lifelength*, F. Proschan and R.J. Serfling (Eds.), Philadelphia: Society for Industrial and Applied Mathematics, 327–358.

2. Chen, Y.Y., Hollander, M. and Langberg, N.A. (1982). Small-sample results for the Kaplan-Meier estimator. *J. Am. Statist. Assoc.* **77**, 141–144.

3. Efron, B. (1967). The two sample problem with censored data. *Proceedings of the Fifth Berkeley Symposium on Mathematical Statistics and Probability* **4**, 831–852.

4. Geurts, J.H.J. (1985). Some small-sample non-proportional hazards results for the Kaplan-Meier estimator. *Statistica Neerlandica* **39**, 1–13.

5. Geurts, J.H.J. (1987). On the small-sample performance of Efron's and of Gill's version of the product limit estimator under nonproportional hazards. *Biometrics* **43**, 683–692.

6. Gill, R.D. (1980). *Censoring and Stochastic Integrals*. Mathematical Centre Tract No. 124. Amsterdam: Mathematisch Centrum.

7. Gillespie, B.W., Gillespie, J.A. and Iglewicz, B. (1992). A comparison of the bias in four versions of the product-limit estimator. *Biometrika* **79**, 149–155.

8. Kaplan, E.L. and Meier, P. (1958). Nonparametric estimation from incomplete observations. *J. Am. Statist. Assoc.* **53**, 457–481.

9. Klein, J.P. (1988). Small sample properties of censored data estimators

of the cumulative hazard rate, survivor function, and estimators of their variance. Research Report 88/7, Statistical Research Unit, University of Copenhagen.

10. Meier, P. (1975). Estimation of a Distribution Function from Incomplete Observations. In *Perspectives in Probability and Statistics*, J. Gani (Ed.), Sheffield, England.: Applied Probability Trust, 67–87.

11. Moeschberger, M.L. and Klein, J.P. (1985). A comparison of several methods of estimating the survival function when there is extreme right censoring. *Biometrics* **41**, 253–259.

Biographies

Brenda Gillespie is an Assistant Professor in the Department of Biostatistics, University of Michigan. She received her Ph.D. in Statistics in 1989 from Temple University in the area of survival analysis. Contact address: Department of Biostatistics, University of Michigan, 109 S. Observatory, Ann Arbor, MI 48109-2029. e-mail: brenda.gillespie@umich.edu.

Justine Uro is a doctoral student in the Department of Biostatistics, University of Michigan. He obtained his B.S. (1987) and M.S. (1990) in Mathematics from the University of the Philippines, and M.S. (1992) in Biostatistics from the University of Michigan. Contact address: same as above. e-mail: justine.uro@umich.edu

ROTATIONAL ENERGY DISPERSIONS FOR VAN DER WAALS MOLECULAR CLUSTERS

Lawrence L. Lohr, Carl H. Huben
University of Michigan, Ann Arbor MI, USA

Abstract

We have obtained analytic expressions, parametric in centrifugal displacement coordinates, which provide exact classical descriptions of the rotational energy dispersions, that is, the dependence of the combined rotational and vibrational energies on the rotational angular momenta, for small molecular clusters bound by van der Waals interactions modeled by pairwise additive (6-12) Lennard-Jones potentials. The clusters considered consist of three (equilateral triangle), four (tetrahedron), and six (octahedron) units and serve as models for small clusters of rare-gas atoms such as argon. This work represents an extension of our recently published study of analytic rotational energy dispersions for diatomic molecules bound by harmonic oscillator, Morse, or Lennard-Jones potentials [J. Mol. Spectrosc. **155**, 205 (1992)].

The **Maple V** mathematical problem-solving system has been employed in numerous ways in this investigation, in obtaining the exact parametric expressions, in obtaining both power series and Padé approximant (rational function) forms useful to molecular spectroscopists, and in making three-dimensional plots of rotational energy surfaces which portray the variations of the combined rotational and potential energies with the directions of the rotational angular momenta with respect to the molecular frames.

The physical properties of the clusters which are addressed using our results include calculation of

Mathematical Computation with Maple V:
Ideas and Applications
Tom Lee, Editor
©1993 Birkhäuser Boston

quartic and higher-order spectroscopic constants, location of rotational instabilities, and characterization of the "cubic" anisotropies for the spherical top clusters A_4 and A_6 which lead to a partial lifting of their rotational energy-level degeneracies. Of particular interest is the result that for each of these cluster types the preferred direction of the rotational angular momentum is parallel to a molecular four-fold axis, leading to reduced symmetries of D_{2d} for tetrahedral A_4 and D_{4h} for octahedral A_6.

I -Introduction

The distortion of a rotating molecule from its equilibrium geometry and its effect upon the rotational energy levels has been recognized for a long time by molecular spectroscopists. Interest in these centrifugal effects, as they are often called, has increased in recent years due to the development of high-resolution spectroscopic techniques and to major advances in the theoretical description of highly excited rotational states of molecules. In a series of studies (Refs.(1-6)) we presented a new approach to centrifugal distortions and their associated rotational energy stabilizations which exploits *ab initio* electronic structure computational methods. This approach is direct, bypassing in its simplest applications the explicit calculation of spectroscopic constants such as vibrational frequencies as this information is implicitly contained in the *ab initio* electronic energy hypersurface. Specifically the method is particularly useful at any computational level for which analytic gradients of potential-energy hypersurface is available. Results were presented in our first study (Ref.(1)) for H_2^+, NH_3, CH_4, BF_3, and SF_6. More detailed studies followed (Refs.(2-4)) for H_2O, O_3, and PH_3, as well as an outline of a generalized extension of the method (Ref.(5)). The procedure is structurally oriented, that is, it focuses on the question of the size and shape of molecules with nonzero rotational angular momentum.

Centrifugal distortion spectroscopic constants are a very useful form of our computational output, providing an important and indispensable basis for comparison to experimental observations, yet their computation is in a way secondary to the main task. Stated differently, our studies are an exploration of molecular energy in those regions of nuclear-coordinate hyperspace which are accessible by centrifugal distortions from the equilibrium geometry.

In the most recent of our studies (Ref.(6)) we confined our attention to diatomic molecules as modeled by harmonic oscillator (HO), Morse oscillator (MO), and Lennard-Jones oscillator (LJO) potential energy functions. By drawing upon the capabilities of the **Maple V** mathematical problem-solving system we obtained closed-form analytic expressions, parametric in the centrifugal displacement, for the dependence of the classical rotational energy upon the rotational energy momentum, that is, the rotational energy dispersion for each of these three model potentials. Further, through power-series reversions of the angular momentum as a function of displacement we obtained Padé approximants for the rotational energy dispersions which approximate well the exact parametric solutions and which may be used for fitting experimental spectroscopic data as they do not display the high angular momentum divergences associated with traditional power series representations of the energy. A particularly interesting result for the diatomic LJO is that the maximum stable centrifugal displacement is only $(5/2)^{1/6}-1 = 0.16499$ in units of the equilibrium separation for the non-rotating molecule; larger separations lead spontaneously to dissociation.

In our present study we continue the use of the **Maple V** system to obtain and characterize analytic descriptions of the rotational energy dispersions for deformable molecules. Specifically we expand the previous study (Ref.(6)) of the Lennard-Jones (LJO) diatomic to include larger clusters formed from atoms (or molecules) bound together by pairwise-additive LJO interactions. Such systems serve as excellent models for real clusters formed from rare-gas atoms or other closed-shell moieties. Indeed there have been many studies in recent years (Refs.(7-20)) describing the structures, stabilities, and dynamics, both rotational and vibrational, of such clusters with a particular focus on those formed from the rare-gas argon (Ar). Perhaps the closest in style to our present study is that of Li and Jellinek (Ref.(13)), who used a simulation program to explore the distortion, isomerization, and fragmentation of the cluster Ar_{13}. Our study is focused on the smaller clusters A_3, A_4, and A_6, where A is an unspecified "particle" having LJO interactions, with most of the results being obtained and characterized analytically by use of the **Maple V** system.

II-Outline of Procedure

For a polyatomic molecule we consider the total electronic energy E_{el} (within the Born-Oppenheimer approximation) as a function of a set of internal nuclear coordinates $Q = \{Q_i\}$, that is,

$$E_{el} = E_{el}(Q_1 Q_2 Q_3 ... Q_n) = E_{el}(Q). \qquad (1)$$

The energy of rotation about the center of mass may be expressed classically in terms of these same coordinates via the moment of inertia tensor $I(Q)$ and the rotational angular momentum J, where $J = I\omega$, with ω being the angular velocity, as

$$E_{rot} = E_{rot}(Q,J) = (1/2)J_t I^{-1} J, \qquad (2)$$

where I^{-1} is the inverse of the matrix I and J_t is the transpose of the vector J. Thus the effective "potential" energy governing the motion of the nuclei is given by the J-dependent sum of E_{el} and E_{rot}, that is, by

$$E_{eff}(Q,J) = E_{el}(Q) + E_{rot}(Q,J), \qquad (3)$$

where the Q dependence of E_{rot} and thus of E_{eff} is via the moment of inertia tensor $I(Q)$.

For a number of purposes we seek extrema in nuclear coordinate space $Q = \{Q_i\}$ of the effective energy E_{eff}. One important purpose is to find the rotational energy dispersion $E_{eff}(J)$ corresponding to the set of minima which classically are continuous with respect to both the magnitude and direction of J in the molecular frame. For such extrema of E_{eff}, not only do we require that $\nabla(E_{eff}) = 0$, but also that the Hessian (second-derivative) matrix have positive eigenvalues. In many cases the rotational energy will depend on only a subset, say m in number, of the total number, say n, of nuclear coordinates $\{Q_i\}$. In such cases there will for a given J m equations of the type $\nabla_i(E_{el} + E_{rot}) = 0$ and $(n - m)$ of the type $\nabla_i E_{el} = 0$. These conditions defines in principle a centrifugal distortion pathway $Q(J)$, although typically we cannot obtain it analytically and thus cannot obtain the desired rotational energy dispersion $E_{eff}(J)$ analytically. However expressing the pathway as $J(Q)$ and substituting in $E_{eff}(Q,J)$ yields $E_{eff}(Q)$, which together with $J(Q)$ defines the dispersion parametrically. Thus in many cases we can represent the dispersion $E_{eff}(J)$ by the pair of analytic relationships $E_{eff}(Q)$ and $J(Q)$. In the remainder of this paper we present such analytic parametric rotational energy dispersions for van der Waals molecular clusters and show how they may be used to extract higher-order (quartic and above) spectroscopic constants.

III - The Lennard-Jones (6-12) Potential and van der Waals Clusters

The molecular systems whose rotational energy dispersions we wish to characterize are the weakly bound clusters of closed-shell molecules known as van der Waals clusters. Typical examples include clusters of rare-gas atoms or clusters of such molecules as carbon dioxide CO_2 or sulfur hexafluoride SF_6. Such clusters, and particularly those comprised of rare-gas atoms, are often modeled by a total electronic energy $E_{el}(Q)$ taken as a sum of pairwise-additive Lennard-Jones (6-12) oscillator (LJO) potential energy terms, where each term is of the form

$$V(r) = 4d[(\sigma/r)^{12} - (\sigma/r)^6]. \quad (4)$$

In the above d is the well depth and σ is the "collision diameter." The attractive (negative) r^{-6} represents the London dispersion energy associated with the interaction of instantaneous electric dipole moments, while the repulsive (positive) r^{-12} term is a convenient representation. The equilibrium separation r_e is given by $2^{1/6}\sigma$. For purposes of this study we find it convenient to re-write Eq. (4) by defining a reduced displacement $x = (r - r_e)/r_e$ and a new variable $z = 1/(1 + x)^2 = (r_e/r)^2$, and by adding d to the energy, giving

$$V(z) = d (1 - z^3)^2. \quad (5)$$

Further, dividing by d yields the dimensionless reduced form

$$v(z) = V(z)/d = (1 - z^3)^2. \quad (6)$$

This form resembles that for the harmonic (Hooke's Law) oscillator, but with the potential energy being proportional to the square of a function of the displacement instead of to the square of the displacement itself. Further, as will be shown, many analytic relationships for pairwise-additive LJO systems have the form of polynomials in the displacement-related parameter z.

While most of our results will be presented and discussed for the dimensionless reduced form (Eq. (6)) for the LJO potential energy, thus making them applicable to any LJO clusters made up of identical moieties, we do present some numerical results for one very important class of examples, namely van der Waals clusters of argon (Ar) atoms. The parameters we have selected for the reference

diatomic Ar_2 are the same as those used by Leitner *et al.* (Ref. (12)) in their study of the cluster Ar_3, namely $\sigma = 3.40$ Å, corresponding to $r_e = 3.82$ Å, and d = 84.1 cm^{-1}. Other sets of Ar_2 parameters have also been proposed or used (Refs.(21-25)). For the isotopic species $^{40}Ar_2$ the rigid-rotor constant $B_e = \hbar^2/2\mu r_e^2 = 0.058$ cm^{-1}, where $\hbar$ is Planck's constant h divided by 2π and μ is the reduced mass $m(^{40}Ar)/2$, so that the dimensionless reduced rigid-rotor constant $\beta = B_e/d = 6.878 \times 10^{-4}$.

IV - Analytic Rotational Energy Dispersions

In Table I we present analytic rotational energy dispersion relationships parametric in centrifugal displacements for a number of simple cases, namely for the LJO clusters A_2, A_3 ($J \parallel C_3$ and $J \parallel C_2$), A_4 ($J \parallel S_4$ and $J \parallel C_3$), and A_6 ($J \parallel C_4$ and $J \parallel C_3$), with the expressions for A_6 being for the simplified model neglecting the three *trans* interactions. For simplicity of notation we denote the dimensionless reduced effective energy $E_{eff}(Q)/d = \epsilon_{eff}(Q)$ as "ϵ", with no subscript. In each case the expressions are in terms of a single structural parameter $z = (1+x)^{-2}$, where x is the reduced displacement, with the tabulation giving the integer coefficients a and b which multiply the basic expressions. Further, the angular momentum J is taken as dimensionless (as a multiple of $\hbar$) throughout this entire section and considered classically, so that J may be aligned with a principal rotation axis of the molecule. If desired the quantity J^2 may be equated to the quantum number expression $J(J + 1)$. For the simple cases in Table I we find:

$$\beta J^2 = a[z^2(1-z^3)] \quad (7)$$

$$\epsilon = b[(1-z^3)(5z^3+1)] \quad (8)$$

The β parameter in Eq.(7) and in Table I is for diatomic A_2 in each case; this enables direct comparisons of the expressions for different sized clusters. In the remainder of this Section we present the corresponding expressions for several cases not representable in terms of a single structural parameter.

For A_3 (D_{3h}) with $J \perp C_3$ and C_2 (the J_y case) the actual molecular symmetry is C_{2v}, with two edges extended and one (parallel to J) slightly compressed. We designate the two structural parameters as $z = (1+x)^{-2}$ and $Z = (1+X)^{-2}$, respectively, where x and X are the corresponding reduced displacements. The dispersion relations may be expressed parametrically in terms of these two variables as follows:

$$\beta J_y^2 = 16z^4(1-z^3)(\frac{1}{z}-\frac{1}{4Z})^2 \qquad (9)$$

$$0 = z^4(1-z^3) + 2Z^4(1-Z^3) \qquad (10)$$

$$\epsilon = 12z^4(1-z^3)(\frac{1}{z}-\frac{1}{4Z}) + 2(1-z^3)^2 \\ + (1-Z^3)^2 \qquad (11)$$

Because of the interdependence in Eq. (10) of the parameters, the energy depends in effect on a single variable.

For A_6 (O_h) symmetry with $\mathbf{J} \parallel C_3$ the actual molecular symmetry is D_{3d}, with the 12 edges being comprised of two sets of 6, described by the parameters z and ζ, and the 3 *trans* distances being equal and described by the parameter Z, with the moment of inertia depending only upon z. The rotational energy dispersion, including these latter interactions, is represented by the following:

$$\beta J^2 = 144z^2(1-z^3) + 72(Z^4/z^2)(1-Z^3) \qquad (12)$$

$$0 = 36\zeta^2(1-\zeta^3) + 18(Z^4/\zeta^2)(1-Z^3) \qquad (13)$$

$$Z = z\zeta/(z+\zeta) \qquad (14)$$

$$\epsilon = \beta J^2 z/4 + 6(1-z^3)^2 + 6(1-\zeta^3)^2 \\ + 3(1-Z^3)^2 \qquad (15)$$

Note that Eqs. (13) and (14) give the interdependencies of the structural parameters, so that the energy again is in effect a function of a single parameter. The effective energy in Eq.(15) has a value of (588/257) for J = 0, for which $z_o = \zeta_o = (264/257)^{1/3}$ and $Z_o = z_o/2$. The value $z_o = (264/257)^{1/3} = 1.00900$ describes the compression ($x_o = -0.00447$) of the 12 edges due to the attractive *trans* interactions. The energy value (588/257) is relative to that for the unattainable structure with all fifteen interacting pairs being at r_e ($z = \zeta = 1$). It is desirable for some purposes to subtract this constant from the energy so that the energy is zero for J = 0. We follow this latter convention in our energy tabulations and all discussion. Alternatively, relative to disso-

ciated atoms the energy at J = 0 is (588/257)-15, or -12.71206. Note also that β in Eqs. (12) and (15) is the value for diatomic A_2, so that rigid-rotor energy is given by $\beta J^2 z_o/4$.

Again for the A_6 cluster but with $\mathbf{J} \parallel C_4$, so that the actual symmetry is D_{4h}, there are 4 edges described by the parameter z, 8 edges by the parameter ζ, two *trans* distances described by z/2, and the unique *trans* distance, that parallel to $\mathbf{J}$, described by Z. The moment of inertia, as for the case of $\mathbf{J} \parallel C_3$, depends only upon z. The rotational energy dispersion is represented by the following:

$$\beta J^2 = 96z^2(1-z^3) + 24(z/2)^2(1-(z/2)^3) \\ - 48(Z^4/z^2)(1-Z^3) \qquad (16)$$

$$0 = 48\zeta^2(1-\zeta^3) + 24(Z^4/\zeta^2)(1-Z^3) \qquad (17)$$

$$Z = z\zeta/(2(2z-\zeta)) \qquad (18)$$

$$\epsilon = \beta J^2 z/4 + 4(1-z^3)^2 + 8(1-\zeta^3)^2 \\ + 2(1-(z/2)^3)^2 + (1-Z^3)^2 \qquad (19)$$

Through the parameter interdependencies the energy is a function of a single variable.

How do we use **Maple V** to solve these sets of equations? Consider for example the last set, Eqs. (16)-(19), describing A_6 with $\mathbf{J} \parallel C_4$. We obtain numerical solutions in the following sequence: first, select a value of Z greater than 1/2 (the corresponding distance is compressed below $2^{1/2}r_e$ by the rotation); second, solve Eq. (17) for ζ; third, solve Eq. (18) for z; fourth, solve Eq.(16) for βJ^2; and fifth, solve Eq. (19) for the energy ϵ. Other sequences are possible as long as we start with an equation in only two variables, taking one of them as independent.

V - Results and Discussion

The key results will be presented and discussed here, with many quantitative details to be left for a future publication (Ref.(26)). We consider first the equilateral triangular complex A_3. Reduced energies ϵ as a function of βJ^2 were found for three cases, namely $\mathbf{J} \parallel C_2$ (J_x), $\mathbf{J} \perp C_2$ (J_y), and $\mathbf{J} \parallel C_3$ (J_z), as based on the expressions in Table I and on

Eqs. (9-11). The **Maple V** system was invaluable in the solution of these and related non-linear equations through the ease with which one can obtain numerical solutions. We note that the ratio ϵ_x/ϵ_y, identically equal to unity for a rigid symmetric top, falls to 0.97738 at the maximum of $\epsilon_x = 1.8$. Similarly the ratio $2\epsilon_z/\epsilon_y$ rises from unity to a value of 1.11812 at the maximum for ϵ_y. These deviations from unity reflect the non-rigidity of the A_3 rotor. In Figure 1 we show the rotational energy surface in **J**-space, that is, the variation of the energy ϵ with respect to the direction of **J** in the molecular frame for some fixed magnitude of **J**. The surface is not cylindrically symmetrical as it would be for a rigid symmetric top, but rather possesses six-fold rotational symmetry (there are three C_2 axes, with clockwise and counterclockwise rotations about each having the same energy, hence there are six equivalent directions in **J**-space). We further note that past the energy maximum of 1.8 for the J_x case that the preferred structure is actually collinear; the triangular molecule "snaps" open!

It is useful to extract (Ref.(1-6)) reduced quartic coefficients δ from these energies, that is, the coefficients of J^4 in traditional power series expansions of rotational energies. We do this not from rotational constants and vibrational frequencies but rather from the "centrifugal stabilization energies," defined as the difference in energy between rigid and non-rigid rotors with the same angular momentum. Specifically the quartic coefficient is taken as the limit as J approaches zero of the stabilization energy divided by J^4; in reduced units of the diatomic A_2 well depth it becomes the limit of the reduced stabilization energy divided by the square of βJ^2. Expressed in these reduced units the diatomic value of δ is $\beta^7/36$; the $J = 0$ limiting values for A_3 are the fractions 1, 1, and 1/12 of this diatomic value for the J_x, J_y, and J_z cases, respectively. The fact that the values are the same for the J_x and J_y cases indicates that the energy differences between these two cases as shown by the rotational energy surface are associated with a higher-order term, namely sextic (J^6), in a power series expansion, corresponding to the six-fold symmetry described above.

We similarly obtained numerical results for the tetrahedral cluster A_4. The two cases are for **J** $\parallel$ C_3 (C_{3v}) and **J** $\parallel$ S_4 (D_{2d}). The striking result is that the latter is energetically preferred for a given angular momentum, that is, this deformable spherical top will make itself into a symmetric top with a preferred S_4 axis. The reduced δ values have $J = 0$ limits of 1/12 and 1/8 in units of the corresponding

value for diatomic A_2; these may be converted (Ref.(1)) to the "spherical" and "cubic" tensor coefficients δ_s and δ_t as shown. The latter has a limiting value of 1/160 in units of the diatomic δ. Although small, this coefficient is the critical measure of the nonspherical effects of centrifugal distortion and is a measure of the associated splittings of the rigid spherical top rotational energy levels. Finally we note for A_4 that the square-planar structure, unstable for zero or low angular momentum with respect to an out-of-plane puckering leading to a tetrahedral structure, actually becomes the stable form past the energy maximum of 3.6 for the **J** $\parallel$ S_4 case; this behavior, in which the tetrahedron "collapses" into a plane, is similar to that described above for A_3 becoming linear past the energy maximum of 1.8 for the J_x case.

The octahedral cluster A_6 differs from the preceding in that there is no arrangement permitting all interacting pairs to be at the diatomic equilibrium separation r_e. The resulting "compression" of the octahedron was discussed in Section IV. Reduced energies were found for three cases, namely **J** $\parallel$ C_3 (D_{3d}), **J** $\parallel$ C_2 (D_{2h}), and **J** $\parallel$ C_4 (D_{4h}). All interactions, including the three *trans* interactions, were included in each case. As with tetrahedral A_4, the case with **J** $\parallel$ C_3 is highest in energy, thus corresponding to eight energy maxima on the rotational energy surface. This result is somewhat surprising, as the molecules CH_4 (tetrahedral) and SF_6 (octahedral) are known (Ref.(27)) to have opposite behaviors; the **J** $\parallel$ C_3 corresponds to rotational energy maxima for CH_4 but minima for SF_6. Thus the reduced quartic coefficient δ_t for A_6 has the same sign as that for A_4, although the magnitude is a factor of seven smaller (0.00093 for A_6 vs 1/160 = 0.00625 for A_4). The **Maple V** system was again a very powerful tool, greatly facilitating the determination of these limiting (as the angular momentum approaches zero) reduced quartic coefficients.

VI - Summary

We have made extensive use of many features of the **Maple V** problem solving system in carrying out a theoretical and computational investigation of the properties of van der Waals cluster molecules. Specifically we have studied the rotational energy dispersions of small clusters bound by pairwise Lennard-Jones interactions; these clusters model those formed from rare-gas atoms or other closed-shell moieties. The dispersion, which represents the variation of the energy of the deformable molecule with respect to the magnitude and direction of its rotational angular momentum, is typically represented by sets of equations parametric in centrifugal displacement coordinates. The **Maple V** system has been used in the derivation of the dispersion equations, in the characterization of their properties, in the numerical evaluation of their solutions, and in the two- and three-dimensional plotting of the numerical results.

Our specific results include quartic spectroscopic constants obtained analytically from the rotational energy dispersion relationships for the clusters A_3, A_4, and A_6. Results are expressed in reduced, dimensionless units and are thus applicable to a wide range of systems, although specific application is made to argon clusters. A striking result is that for both A_4 and A_6 the sign of the cubic anisotropy in the dispersions is such that these nominally spherical tops each display a preference for the rotational angular momentum to be along a four-fold rotation axis, with angular momentum along a three-fold axis representing a rotational energy maximum.

References

1. L. L. Lohr and J.-M. J. Popa, J. Chem. Phys. **84**, 4196 (1986).

2. L. L. Lohr and A. J. Helman, J. Comput. Chem. **8**, 307 (1987).

3. L. L. Lohr, Int. J. Quantum Chem: Quantum Symp. Symp. **21**, 407 (1987).

4. A. Taleb-Bendiab and L. L. Lohr, J. Mol. Spectrosc. **132**, 413 (1988).

5. L. L. Lohr, J. Mol. Struct. (THEOCHEM) **199**, 265 (1989).

6. L. L. Lohr, J. Mol Spectrosc. **155**, 205 (1992).

7. R. S. Berry, J. Jellinek, and G. Natanson, Chem. Phys. Letters **107**, 227 (1984).

8. R. S. Berry, J. Jellinek, and G. Natanson, Phys. Rev. **A30**, 919 (1984).

9. J. Jellinek, T. L. Beck, and R. S. Berry, J. Chem. Phys. **84**, 278 (1986).

10. R. S. Berry, T. L. Beck,, H. L. Davis, and J. Jellinek, Adv. Chem. Phys. **70 (part 2)**, 75 (1988).

11. N. Quirke, Mol. Simul. **1**, 249 (1988).

12. D. M. Leitner, R. M. Whitnell, and R. S. Berry, J. Chem. Phys. **91**, 3470 (1989).

13. D. H. Li and J. Jellinek, Z. Phys. D - Atoms, Molecules, and Clusters **12**, 177 (1989).

14. J. Jellinek and D. H. Li, Phys. Rev. Letters **62**, 241 (1989).

15. J. Jellinek and D. H. Li, Chem. Phys. Letters **169**, 380 (1990).

16. T. R. Horn, R. B. Gerber, J. J. Valentini, and M. A. Ratner, J. Chem. Phys. **94**, 6728 (1991).

17. H. Cheng and R. S. Berry, Phys. Rev. **B45**, 7969 (1992).

18. D. M. Leitner, J. D. Doll, and R. M. Whitnell, J. Chem. Phys. **96**, 9239 (1992).

19. C. D. Maranas and C. A. Floudas, J. Chem. Phys. **97**, 7667 (1992).

20. X. Hu and W. L. Hase, J. Phys. Chem. **96**, 7535 (1992).

21. R. A. Aziz and H. H. Chen, J. Chem. Phys. **67**, 5719 (1977).

22. E. Scoles, Annu. Rev. Phys. Chem. **31**, 81 (1980).

23. M. J. Ondrechen, Z. Berkovitch-Yellin, and J. Jortner, J. Am. Chem. Soc. **103**, 6586 (1981).

24. C. Douketis, G. Scoles, S. Marchetti, M. Zen, and A. L. Thakker, J. Chem. Phys. **76**, 305 (1982).

25. R. A. Aziz and M. Salman, Mol. Phys. **58**, 679 (1986).

26. L. L. Lohr and C. H. Huben, to be published.

27. For a review see G. Natanson, Rev. Mod. Phys., in press.

Table I

Energy-Angular Momentum Expressions Parametric in Displacements for LJO Clusters

Species	Direction	βJ^2 $[z^2(1-z^3)]^{a,b}$	ϵ $[(1-z^3)(5z^3+1)]^c$
A_2	$J \perp C_\infty$	6^d	1^d
A_3 (D_{3h})			
	$J \parallel C_3$	36	3
	$J \parallel C_2$	6	1
A_4 (T_d)			
	$J \parallel S_4$	24	2
	$J \parallel C_3$	36	3
A_6 (O_h - *cis* only)			
	$J \parallel C_4$	96	4
	$J \parallel C_3$	144	6

a) $z = 1/(1 + x)^2 = (r_e/r)^2$

b) $\beta = B_e(A_2)/D_e(A_2)$, where $D_e(A_2) =$ well depth d.

c) $\epsilon = E_{eff}(Q)/D_e(A_2)$.

d)   The tabulated integers are multipliers of the expressions at the head of each column.

Legend for Figure

1.   Rotational energy surface for the LJO cluster A_3. While the (undistorted) molecular symmetry is D_{3h}, that for the surface is D_{6h}. The upper figure (a) is to scale based on energies of 1.8, 1.84166, and 0.96800 (units of diatomic A_2 dissociation energy) for J_x, J_y, and J_z, respectively ($\beta J^2 = 9(2/5)^{5/3} = 1.95438$) and is obtained by interpolations assuming $\cos(6\phi)$ and $\cos(2\theta)$ dependences. The lower figure (b) has greatly exaggerated angular variations for emphasis.

a)
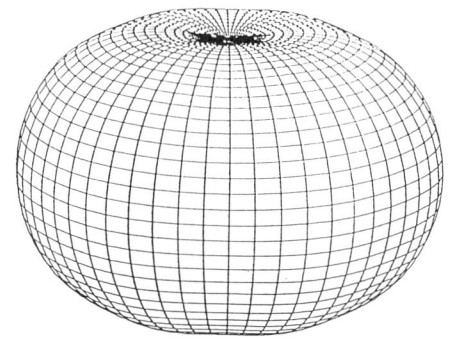

b)
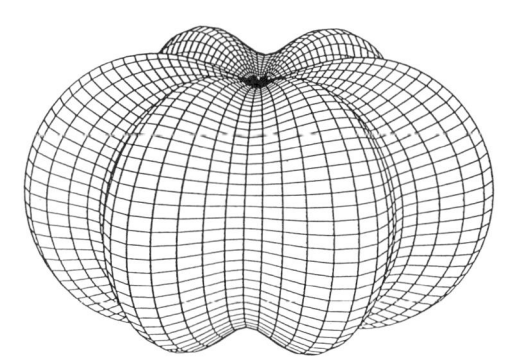

Biographies of Authors

Lawrence L. Lohr, Professor of Chemistry, has been on the faculty of the University of Michigan, Ann Arbor, since 1968. He is spending the 1992-93 academic year as the Program Officer for Theoretical and Computational Chemistry at the National Science Foundation, Washington, DC. He received a B.S. Chemistry degree from the University of North Carolina, Chapel Hill, in 1959, and a Ph.D. in physical chemistry from Harvard University in 1964. He spent two years as a Research Associate at the University of Chicago and three years as a Research Scientist at the Ford Motor Co. Scientific Laboratory (Dearborn, MI) before joining the University of Michigan faculty. He has spent extended periods at Bell Telephone Laboratories (Murray Hill, NJ), The University of California (Berkeley), The University of Georgia (Athens), The Institute for Molecular Science (Okazaki, Japan), and The University of Helsinki (Finland). He is the author of more than one hundred journal articles on various aspects of both theoretical and experimental physical chemistry.

Professor Lohr may be reached at The Department of Chemistry, The University of Michigan, Ann Arbor, MI, 48109-1055. His e-mail address is: lawrence.lohr@um.cc.umich.edu.

Carl H. Huben was graduated with a B.S. Chemical Engineering degree from the University of Michigan, Ann Arbor, in December, 1992. A recipient of a National Science Foundation Predoctoral Fellowship, he plans to pursue graduate studies in physical chemistry at Stanford University.

Mr. Huben may be reached c/o Professor Lawrence L. Lohr at The Department of Chemistry, The University of Michigan, Ann Arbor, MI, 48109-1055. Mr. Huben's e-mail address is: carl.huben@um.cc.umich.edu.

SYMBOLIC COMPUATION IN COMPUTABLE GENERAL EQUILIBRIUM MODELING

Trien T. Nguyen
Department of Economics, University of Waterloo, Waterloo ON, Canada

1   Introduction

Alongside theoretical developments of mathematical economics in the fifties and sixties, the emerging field of computable general equilibrium (CGE) modeling (Scarf and Shoven [1984], Shoven and Whalley [1992]) has been pre-occupied with the problem of numerically solving systems of highly nonlinear simultaneous equations describing general equilibrium states of a multi-sector model of an economy. Path breaking contributions by Scarf [1973], Merrill [1972], van der Laan and Talman [1979], and others in the theory of fixed-point computation have made it now possible to routinely compute numerical solutions of complex CGE models in various areas such as public finance, international trade, development, environment, and economic history.

The availability of modeling software such as GAMS (General Algebraic Modeling System) (see Brooke et al. [1992], Meeraus [1983]), originally developed at the World Bank, has allowed modelers to concentrate on economic modeling issues while leaving numerical computational aspects to solvers such as MINOS (Murtagh and Saunders [1987]). As more modelers in government agencies and research institutions are using modeling software for policy analysis, it is much less often that modelers have to write their own solvers in higher programming languages such as FORTRAN, Pascal or C.

While GAMS allows modelers to concentrate on modeling issues, it is however in this modeling part

Mathematical Computation with Maple V:
Ideas and Applications
Tom Lee, Editor

that they are completely left unassisted. They still have to mainly rely on the traditional "paper and pencil" technology in the specification of their models. Analytical expressions for specific demand and production functional forms such as CES (constant elasticity of substitution), VES (variable elasticity of substitution), or LES (linear expenditure system), with or without multi-level nesting, remain to be worked out *manually* by modelers. Calibration procedures (Mansur and Whalley [1984]), no matter how laborious and tedious, still have to be performed manually so that analytical formulas for key model parameters can be derived before the solver in GAMS can crank out their numerical values. If modelers wish to get a more realistic representation of the real world, the modeling process will become so complicated that the analytical part can quickly render the model intractable by hand even before the numerical part has a chance to tackle its job.

Symbolic computation software such as Maple (Char et al. [1991ab]) can be used to assist modelers in the process of model building. In this paper, we present an *interface* approach in which Maple can be used as a front-end preprocessor generating outputs in GAMS format which can be later processed by GAMS for large-scale numerical calculations. Conversely, though not considered in the paper, it is also conceivable that GAMS (using the PUT facility) can be used as a back-end postprocessor to generate outputs in Maple format which can be processed by Maple for post-optimal analysis such as sensitivity, stability, multiplicity, and graphical presentations. That is, a two-way communication can be made between these two powerful software tools for the benefits of modelers.

The organization of the paper is as follows: section 2 provides the theoretical background for a simple illustrative example in CGE modeling. Section 3 shows how the example is numerically solved using a *manually-coded* GAMS program. This section also serves as an exposition on the structure of GAMS programs for readers who are not familiar with this modeling tool. Section 4 demonstrates how Maple can contribute to the modeling process with Maple-*assisted* GAMS version of the program. Section 5 concludes the paper with a brief summary and conclusions.

2   A Simple Economic Model

Consider the following simple illustrative model of an economy with n commodities indexed by $j = 1, \ldots, n$ and h households indexed by $i = 1, \ldots, h$. Each household is given fixed amounts of commodities called *initial endowments* $W_{ij} > 0$ and is allowed to trade in the market for the "right" amounts of commodities to their own satisfaction. Satisfaction is described by the so-called *utility function* $U_i (X_{i1}, \ldots, X_{in})$ which can be specified by either one of the following two popular functional forms:

$$
U_i \;=\; \begin{cases} \displaystyle\prod_{j=1}^{n} X_{ij}{}^{\delta_{ij}} & \text{(Cobb-Douglas)} \\[2em] \left[\displaystyle\sum_{j=1}^{n} \delta_{ij}\, X_{ij}{}^{\rho_i} \right]^{\frac{1}{\rho_i}} & \text{(CES)} \end{cases} \quad (1)
$$

with share parameters $\delta_{ij} > 0$, *elasticity of substitution* $\sigma_i \geq 0$, $\rho_i = 1 - (1/\sigma_i)$, and quantities demanded X_{ij}. The value of σ_i indicates how easily commodities can be substituted for each other to maintain household satisfaction. For example, $\sigma_i = \infty$ implies infinite or perfect substitutability while $\sigma_i = 0$ means no substitution at all, for example, like right shoes and left shoes.

Given any price vector $(p_1, \ldots, p_n)$, *endowment incomes* Y_i can be evaluated as the value of initial endowments at current market prices:

$$
Y_i (p_1, \ldots, p_n) \;=\; \sum_{j=1}^{n} p_j\, W_{ij}. \quad (2)
$$

Endowment incomes enable households to buy the maximum quantities of commodities which they can afford. The economic problem for households in this simple framework is therefore to balance between what they want (satisfaction) and what they can afford (income constraint). This can be

mathematically formulated as the following classical constrained optimization problem:

$$
\begin{aligned} \text{maximize} \quad & U_i (X_{i1}, \ldots, X_{in}) \\ \text{subject to} \quad & \sum_{j=1}^{n} p_j\, X_{ij} = Y_i \end{aligned} \quad (3)
$$

which can be solved either manually or with Maple (see subsection 4.2) for household demands $X_{i1}, \ldots, X_{in}$:

$$
X_{ij} \;=\; \begin{cases} \dfrac{\delta_{ij}}{p_j}\, Y_i & \text{(Cobb-Douglas)} \\[2.5em] \dfrac{\left[\dfrac{\delta_{ij}}{p_j} \right]^{\sigma_i} Y_i}{\displaystyle\sum_{k=1}^{n} \delta_{ik}{}^{\sigma_i}\, p_k^{(1-\sigma_i)}} & \text{(CES)} \end{cases} \quad (4)
$$

Clearly, quantities demanded are negatively related to prices and positively related to incomes. This is consistent with intuition since, everything else being the same, the lower the prices or the higher the incomes, the greater quantities people will buy. Individual excess demands are defined as differences between quantities demanded and initial endowments:

$$
Z_{ij} (p_1, \ldots, p_n) \;=\; X_{ij} - W_{ij}. \quad (5)
$$

A positive individual excess demand means that the household is buying to make up for the gap $X_{ij} - W_{ij}$ while a negative individual excess demand means that the household is selling the excess supply $W_{ij} - X_{ij}$. A zero individual excess demand means that the household is neither buying nor selling since quantity demanded matches exactly with the initial endowment.

Market excess demands are defined as the sum of all individual excess demands Z_{ij} in (5):

$$
Z_j (p_1, \ldots, p_n) \;=\; \sum_{i=1}^{h} (X_{ij} - W_{ij}) \quad (6)
$$

A positive value for Z_j means that the economy as a whole is having shortages since households altogether demand more than the total amount of endowments in the economy. On the other hand, a negative value for Z_j implies surpluses since households demand less than the amount of endowments available. The economy is said to be in *general equilibrium* when quantities demanded match exactly with quantities supplied for all commodities,

$$
Z_j (p_1, \ldots, p_n) \;=\; 0 \quad (7)
$$

in which case the corresponding price vector is called an *equilibrium price vector*. The existence proof for such an equilibrium price vector is a non-trivial one requiring fixed-point arguments, usually in the form of fixed-point theorems by Brouwer or Kakutani (see Debreu [1959], Weintraub [1983]). Much of the development of modern mathematical economics during the last four decades has been centering around this concept of general equilibrium.

3 Anatomy of GAMS

This section outlines the structure of GAMS programs in the context of the illustrative example presented in the previous section. To be specific, suppose that there are three commodities ($n = 3$) named C1, C2, C3 and two households ($h = 2$) named H1, H2. Model parameters such as elasticities of substitution σ_i, share parameters δ_{ij}, and initial endowments W_{ij} are given in the table below.

Elasticity of Substitution		Share Parameters			Initial Endowments		
		C1	C2	C3	C1	C2	C3
H1	1	2/9	1/3	4/9	2	3	4
H2	2	1/4	1/3	5/12	3	4	5

Because GAMS follows the algebraic representation of the model as closely as possible, its program structure is quite intuitive and transparent. GAMS statements are in free format with a maximum line length of 120 characters. Comment are marked by an asterisk (∗) in the first column and inline comments are user-definable. Unlike Maple, GAMS is not case sensitive. Separators are commas, semi-colons, and newline characters but relaxed punctuation is permitted.

Subsection 3.1 presents the GAMS data file `gexparam.inc` which specifies necessary parameters of the model while subsection 3.2 presents the main program `gexgams.gms` which numerically solves the model using a nonlinear programming method. The main program was *manually coded* as usually the case of actual modeling work.

3.1 Parameter Specification

The data file `gexparam.inc` listed below declares in GAMS format various model parameters for our simple illustrative example. The file consists of two declarations: model dimension and parameter specification. For model dimension, the **SETS** statement defines two index sets, namely, $j = \{C1, C2, C3\}$ for commodities and $i = \{H1, H2\}$ for households. For parameter specification, the **PARAMETERS** statement declares elasticities of substitution `sigma(i)`, share parameters `delta(i,j)`, and initial endowments `W(i,j)`.

```
*****  GAMS FILE gexparam.inc  *****
SETS
   j  commodity  /  C1,  C2,  C3  /
   i  household  /  H1,  H2  /;

PARAMETERS
   sigma(i)   elasticity of substitution
      /
        H1   1.
        H2   2.
      /
   delta(i,j)   share parameters
      /
        H1.C1   .2222222222
        H1.C2   .3333333333
        H1.C3   .4444444444

        H2.C1   .2500000000
        H2.C2   .3333333333
        H2.C3   .4166666667
      /
   W(i,j)   initial endowments
      /
        H1.C1   2.
        H1.C2   3.
        H1.C3   4.

        H2.C1   3.
        H2.C2   4.
        H2.C3   5.
      /;
```

3.2 The Manually-Coded GAMS Solution

The *manually coded* GAMS program `gexgams.gms` follows closely, almost in a step-by-step fashion, the theoretical framework in section 2. The program consists of five building blocks as follows:

1. The *parameter block* reads the data file `gexparam.inc` with the `$INCLUDE` statement. Model dimension and parameters are declared in this block.

2. The *variable block* has two variable declaration statements. The `POSITIVE VARIABLES` statement declares variables which accept only *nonnegative* values such as prices `p(j)`, endowment incomes `I(i)`, and household demands `X(i,j)`. On the other hand, the `FREE VARIABLES` statement declares *unbounded* variables such as market excess demands `Z(j)` and objective function `object`.

3. The *equation block* spells out the algebraic structure of the model and the objective function with four `EQUATIONS` statements: `EQ_Y(i)` for endowment incomes `Y(i)` as in (2), `EQ_X(i,j)` for household demands `X(i,j)` as in (4), `EQ_Z(j)` for market excess demands `Z(j)` as in (6), and `EQ_object` for the objective function $\texttt{object} = \sum_{j=1}^{n}\left(p_j Z_j\right)^2 \geq 0$.

   It is this equation block that computer algebra software such as Maple can assist modelers in the problem-solving process. The following features of GAMS syntax are noteworthy: (i) the `ALIAS(j,k)` statement defines `k` as an additional running index equivalent to `j`; (ii) summations are written as `sum(j,...)` in contrast with `sum(...,j=1..n)` in Maple; (iii) the equal sign in equation statements is denoted by "`=E=`" in contrast with the usual equal sign "`=`" in assignment statements.

4. The *model block* consists of, in this simple example, a single `MODEL` statement which in effect defines the model name as `gex` (general equilibrium in exchange). The specification `/ALL/` means that *all* equations `EQ_Y(i)`, `EQ_X(i,j)`, `EQ_Z(j)`, and `EQ_object` of the equation block, are used to describe the model.

5. The *solve block* consists of two parts: (i) initializing variables by setting *lower bounds* using `.LO` suffixes for positive variables (0.000001 for prices to avoid division by zero) and setting *levels* using `.L` suffixes for relevant variables; and (ii) solving the model with the `SOLVE` statement which in effect instructs GAMS to minimize the objective function using a nonlinear programming solver such as MINOS.

The objective function `object` basically reflects the sum of squared deviations of the value of all market excess demands Z_j evaluated at current market prices. An optimal solution at which this objective function is minimized, therefore, can be viewed, for practical purposes, as an approximation of a general equilibrium with zero market excess demands.

Listing of the *manually-coded* GAMS program `gexgams.gms` is as follows:

```
*****   GAMS FILE  gexgams.gms   *****
*****   manually-coded           *****

$INCLUDE "gexparam.inc"

POSITIVE VARIABLES
   p(j)        prices
   Y(h)        endowment incomes
   X(h,j)      household demands;

FREE VARIABLES
   Z(j)        market excess demands
   object      objective function;

* -------------------------------------
EQUATIONS
   EQ_Y(h)    endowment incomes (2)
   EQ_X(h,j)  household demands (4)
   EQ_Z(j)    market excess demands (6)
   EQ_object  objective function;

   ALIAS(j,k);

   EQ_Y(h)..  Y(h) =E=
      sum(j,p(j) * W(h,j));

   EQ_X(h,j).. X(h,j) =E=
      ((delta(h,j)/p(j))**sigma(h)) *
      Y(h) /
      sum(k,(delta(h,k)**sigma(h)) *
           (p(k)**(1.0-sigma(h))));

   EQ_Z(j).. Z(j) =E=
      sum(h,X(h,j) - W(h,j));

   EQ_object.. object =E=
      sum(j,(p(j) * Z(j))**2);
* -----------------------------------

MODEL gex /ALL/;

Y.LO(h)   = 0;
```

```
X.LO(h,j) = 0;
p.LO(j)  = 0.000001;
p.L(j)   = 1.0;
Z.L(j)   = 0;
```

```
SOLVE gex MINIMIZING object USING NLP;
```

GAMS produces the following general equilibrium solution which shows that household H1 buys C1, C2 and sells C3 while household H2 sells C1, C2 and buys C3. Speaking of the economy as a whole, quantities demanded by both households match exactly with total endowments available. In other words, market excess demands are zero for all commodities as required by theory.

	Commodities		
	C1	C2	C3
Equilibrium Prices	1.0000	1.1108	1.2045
Household Demands			
H1	2.2556	3.0459	3.7454
H2	2.7444	3.9541	5.2546
Total	5.0000	7.0000	9.0000
Initial Endowments			
H1	2.0000	3.0000	4.0000
H2	3.0000	4.0000	5.0000
Total	5.0000	7.0000	9.0000
Market Excess Demands	0.0000	0.0000	0.0000

4 Interfacing GAMS with Maple

Section 3 shows how the gex model was numerically solved using GAMS alone. For this simple illustrative example, an argument can certainly be made that numerical software like GAMS provides modelers with all the tools they need. This is however based on the presumption that the analytical structure of the model, especially equation (4) of household demands, has been manually worked out in full and correctly. Given this analytical structure, the remaining task is to code it into proper GAMS format before specifying a nonlinear programming solver such as MINOS for numerical solutions. No measurements of time costs to modelers have been recorded for this tedious and error-prone manual process of analytical derivation of the equation block required by GAMS. In this section, we present an *interface* approach (see figure 1) in which tools in symbolic computation such as Maple can be successfully used in conjunction with modeling software like GAMS. The general idea is to use Maple as a front-end processor generating outputs in GAMS format which can be later processed by GAMS for numerical calculations.

4.1 Parameter Block

The Maple program gexparam.mpl listed below has two parts. The first part specifies model parameters selected in the previous section such as n commodities, h households, elasticities of substitution sigma (all Cobb-Douglas except the last household), utility share parameters delta, and initial endowments W. For illustration purposes, these data are purely hypothetical with no real economic significance. The second part of the program writes these model parameters back into a parameter block in the GAMS data file gexparam.inc listed in subsection 3.1 above.

```
#####  Maple file gexparam.mpl  #####
#####  included by gexmaple.mpl #####

#####  global parameters  #####
n := 3;
h := 2;
sigma := array(1..h,[seq(1,i=1..h-1),h]);
delta := array(1..h,1..n);
W     := array(1..h,1..n);
for i from 1 to h do
   for j from 1 to n do
      delta[i,j] :=
         (i+j)/sum(i+jj,jj=1..n)
   od;
   for j from 1 to n do
      W[i,j] := (i+j)
   od;
od;

#####  parameter block  #####
interface(quiet=true);
```

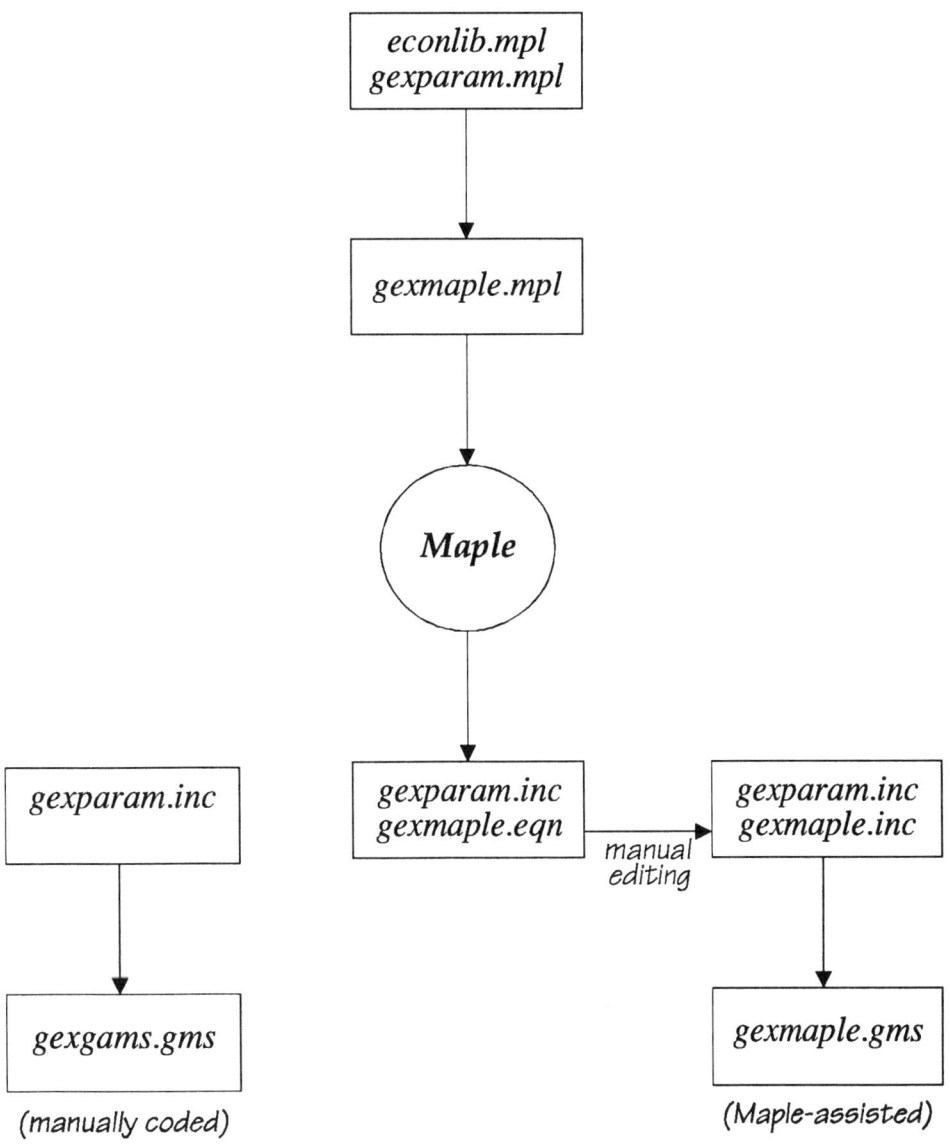

Fig. 1. Interfacing GAMS with Maple

```
writeto('gexparam.inc');
lprint('*****  GAMS FILE  ',
       'gexparam.inc  *****');
lprint('SETS');
lprint('    j   commodity  /',
       seq(cat('C',jj,','),jj=1..n-1),
       cat('C',n),'/');
lprint('    i   household  /',
       seq(cat('H',ii,','),ii=1..h-1),
       cat('H',h),'/;');
lprint();
lprint('PARAMETERS');
lprint('    sigma(i)  ',
       'elasticity of substitution');
lprint('         /');
for i from 1 to h do
   lprint('        ',
          cat('H',i),
          evalf(sigma[i]))
   od;
lprint('         /');
lprint('    delta(i,j)',
       'share parameters');
lprint('         /');

for i from 1 to h do
   for j to n do
      lprint('         ',
             cat('H',i,'.C',j),
             evalf(delta[i,j]))
      od;
   if i < h then lprint() fi;
od;
lprint('         /');
lprint('    W(h,j)  ',
       'initial endowments');
lprint('         /');
for i from 1 to h do
   for j to n do
      lprint('         ',
             cat('H',i,'.C',j),
             evalf(W[i,j]))
      od;
   if i < h then lprint() fi;
od;
lprint('         /;');
```

4.2 Household Demands

The Maple file `econlib.mpl` listed below consists of two procedures. The procedure `cdces` is a direct translation of the utility function U_i (1) into Maple format. The procedure `umax` solves the util-

ity maximization problem (3) for household demands X_{ij}. Using the Lagrangian method, the problem is transformed into one of maximizing the Lagrangian $\mathcal{L} = U_i(X_{i1}, \ldots, X_{in}) + \lambda(Y_i - \sum_{j=1}^{n} p_j X_{ij})$ and solving the following first-order conditions:

$$mrs_{jk} \equiv \frac{\left[\frac{\partial U_i}{\partial X_{ij}}\right]}{\left[\frac{\partial U_i}{\partial X_{ik}}\right]} = \frac{\delta_{ij}\, X_{ij}^{-(\frac{1}{\sigma_i})}}{\delta_{ik}\, X_{ik}^{-(\frac{1}{\sigma_i})}} = \frac{p_j}{p_k} \quad (8)$$

$$\sum_{j=1}^{n} p_j X_{ij} = Y_i \quad (9)$$

In economic jargon, mrs_{jk} is called the *marginal rate of substitution* to describe the extent to which commodities j and k can be marginally substituted for each other without jeopardizing household utility. The exponential term $-(1/\sigma_i)$ on the unknown variable X_{ij} in equation (8) is the main source of nonlinearity which makes a direct application of Maple's built-in **solve** routine unsuccessful. However, a bit of human touch by raising both sides of equation (8) to the exponential power of σ_i

$$\frac{\delta_{ij}^{\sigma_i}\, X_{ij}}{\delta_{ik}^{\sigma_i}\, X_{ik}} = \frac{p_j^{\sigma_i}}{p_k^{\sigma_i}} \quad (10)$$

will remove the nonlinearity and the built-in **solve** routine now can be successfully applied to the linear system (10, 9). This suggests that the powerful mathematical capability of Maple should be meant to complement rather than substitute modelers in the problem-solving process. Listing of the Maple program `econlib.mpl` is as follows:

```
#####  Maple file econlib.mpl  #####
#####  included by gexmaple.mpl #####

#####  utility function in (1)  #####
cdces := proc(sigma,n,delta,x)
   local j, rho;
   if sigma=1 then
      product(x[j]^delta[j],j=1..n)
   elif sigma<>0 then
      rho := 1 - (1/sigma);
      (sum((delta[j]*(x[j]^rho)),
                       j=1..n))^(1/rho)
   else
      ERROR('zero sigma')
   fi;
end:
```

```
##### household demands in (10,9) #####
umax := proc(umax_u,s,n,
            umax_p,umax_i,umax_x)
  local i, j,
        mu, mrs,
        ce, eqnset, varset;
  for j from 1 to n do
    mu[j] := simplify(
             diff(umax_u,umax_x[j]))
    od:
  mrs := array(1..n,1..n);
  for i from 1 to n do
    for j from 1 to n do
      # see equation (8)
      mrs[i,j] := simplify(
                  mu[i]/mu[j]);
      if s=1 then
        ce[i] := mrs[i,j] =
                 umax_p[i]/umax_p[j];
      else
        lhs := simplify(
               mrs[i,j]^(-s),power);
        rhs := (umax_p[i]/
               umax_p[j])^(-s);
        # see equation (10)
        ce[i]  := lhs = rhs;
      fi;
    od;
  od;
  # see equation (9)
  ce[n] := sum(umax_p[k]*umax_x[k],
           k=1..n) = umax_i;
  eqnset := convert(ce,set);
  varset := convert(umax_x,set);
  assign(solve(eqnset,varset));
end:
```

Maple returns the following demands by household
H1 with Cobb-Douglas utility function:

$$U[1] := X[1,1]^{2/9} \; X[1,2]^{1/3} \; X[1,3]^{4/9}$$

$$X[1,1] := \frac{2}{9} \; \frac{2 \; p[1] + 3 \; p[2] + 4 \; p[3]}{p[1]}$$

$$X[1,2] := \frac{1}{3} \; \frac{2 \; p[1] + 3 \; p[2] + 4 \; p[3]}{p[2]}$$

$$X[1,3] := \frac{4}{9} \; \frac{2 \; p[1] + 3 \; p[2] + 4 \; p[3]}{p[3]}$$

and household H2 with CES utility function:

$$U[2] := \left(\frac{1}{4} X[2,1]^{1/2} + \frac{1}{3} X[2,2]^{1/2} + \frac{5}{12} X[2,3]^{1/2} \right)^2$$

$$X[2,1] := 9 \; \frac{p[3] \; p[2] \; DEN}{p[1] \; NUM}$$

$$X[2,2] := 16 \; \frac{p[3] \; p[1] \; DEN}{p[2] \; NUM}$$

$$X[2,3] := 25 \; \frac{p[2] \; p[1] \; DEN}{p[3] \; NUM}$$

```
NUM = (9 p[3] p[2] + 16 p[3] p[1] +
                25 p[2] p[1])

DEN = (3 p[1] + 4 p[2] + 5 p[3])
```

4.3 Market Excess Demands

The Maple interface program gexmaple.mpl first
reads associated files econlib.mpl, gexparam.mpl
and then proceeds to calculate market excess demands in the following order: utility functions as
in routine cdces, endowment incomes as in equation (2), utility maximization as in routine umax,
household excess demands as in equation (5), and
finally market excess demands as in equation (6).

```
##### Maple file gexmaple.mpl #####

##### read included files #####
read('econlib.mpl'): # economic routines
read('gexparam.mpl'): # model parameters

##### global variables #####
p   := array(1..n);      # prices
U   := array(1..h);      # utility
Y   := array(1..h);      # incomes
X   := array(1..h,1..n); # demands
Zij := array(1..h,1..n); # excess demands
Z   := array(1..n);      # excess demands

##### market excess demands #####
for i from 1 to h do
```

```
# defined in econlib.mpl
U[i] := cdces(sigma[i],n,
              [seq(delta[i,jj],jj=1..n)],
              [seq(X[i,jj],jj=1..n)]
              );
# see equation (2)
Y[i] := sum(p[k]*W[i,k],k=1..n);
# defined in econlib.mpl
umax(U[i],
     sigma[i],n,p,Y[i],
     [seq(X[i,jj],jj=1..n)]
     );
for j from 1 to n do
   # see equation (5)
   Zij[i,j] := eval(X[i,j]) - W[i,j]
od;
od;
for j from 1 to n do
   # see equation (6)
   Z[j] := sum(Zij[k,j], k=1..h);
   Z[j] := simplify(Z[j]);
od;
```

Maple returns market excess demands as a system of nonlinear equations in terms of prices alone as follows:

$$Z[1] := -1/9 \, (-170 \, p[3] \, p[1] \, p[2] +$$
$$656 \, p[3] \, p[1]^2 + 1025 \, p[1] \, p[2]^2 -$$
$$378 \, p[3] \, p[2]^2 - 150 \, p[2] \, p[1]^2 -$$
$$477 \, p[3]^2 \, p[2] - 128 \, p[3]^2 \, p[1])/$$
$$(p[1] \, (9 \, p[3] \, p[2] + 16 \, p[3] \, p[1] +$$
$$25 \, p[2] \, p[1]))$$

$$Z[2] := 2/3 \, (11 \, p[3] \, p[1] \, p[2] +$$
$$88 \, p[3] \, p[1]^2 + 25 \, p[1] \, p[2]^2 -$$
$$81 \, p[3] \, p[2]^2 - 225 \, p[2] \, p[1]^2 +$$
$$18 \, p[3]^2 \, p[2] + 152 \, p[3]^2 \, p[1])/$$
$$(p[2] \, (9 \, p[3] \, p[2] + 16 \, p[3] \, p[1] +$$
$$25 \, p[2] \, p[1]))$$

$$Z[3] := 1/9 \, (-236 \, p[3] \, p[1] \, p[2] +$$
$$128 \, p[3] \, p[1]^2 + 875 \, p[1] \, p[2]^2 +$$
$$108 \, p[3] \, p[2]^2 + 1200 \, p[2]^2 \, p[1] -$$
$$585 \, p[3]^2 \, p[2] - 1040 \, p[3]^2 \, p[1])/$$
$$(p[3] \, (9 \, p[3] \, p[2] + 16 \, p[3] \, p[1] +$$
$$25 \, p[2] \, p[1]))$$

Both the Maple program `gexmaple.mpl` of this section and the manually-coded GAMS program `gexgams.gms` of the previous section follow closely the theoretical framework outlined in section 2. However, while the GAMS program builds its calculations of market excess demands based on previously defined variables such as endowment incomes, household demands and excess demands, the Maple program eliminates all intermediate values and expresses market excess demands as final expressions in terms of prices alone. Although this feature of Maple can easily produce unwieldy messy outputs, it clearly shows the basic tenet of economic theory, namely, *prices alone are the key driving force of a market economy.*

4.4 Equation Block

The remaining portion of the Maple interface program `gexmaple.mpl` is basically a GAMS generator which writes the Maple outputs of market excess demand functions in suitable GAMS format.

```
##### Maple file gexmaple.mpl  #####
##### GAMS program generator   #####

##### write GAMS equation block #####
interface(quiet=true);
writeto('gexmaple.eqn');
lprint('*****  GAMS FILE ',
       'gexmaple.eqn  *****');
lprint('EQUATIONS');
for j from 1 to n do
   lprint(cat('   EQ_Z',j,
          '  market excess demand'))
od;
lprint();
lprint('   EQ_object  ',
       'objective function;');
lprint();
lprint('   EQ_object..  ',
       'object =E= ');
lprint('sum(j,(p(j) * Z(j)) ** 2);');
lprint();
```

```
for j from 1 to n do
   lprint(cat(`    EQ_Z`,j,
              `.. Z("C`,j,`") =E= `));
   lprint(Z[j],`;`);
   lprint();
od;
```

Maple returns the output file `gexmaple.eqn` in GAMS format as follows:

```
*****  GAMS FILE  gexmaple.eqn  *****
EQUATIONS
    EQ_Z1   market excess demand
    EQ_Z2   market excess demand
    EQ_Z3   market excess demand
    EQ_object  objective function;

    EQ_object.. object =E=
sum(j,(p(j) * Z(j)) ** 2);

    EQ_Z1.. Z("C1") =E=
-1/9*(-170*p[3]*p[1]*p[2]+656*p[3]*
p[1]**2+1025*p[1]**2*p[2]-378*p[3]*
p[2]**2-150*p[2]**2*p[1]-477*
p[3]**2*p[2]-128*p[3]**2*p[1])/
p[1]/(9*p[3]*p[2]+16*p[3]*p[1]+
25*p[2]*p[1])    ;

    EQ_Z2.. Z("C2") =E=
2/3*(11*p[3]*p[1]*p[2]+88*p[3]*
p[1]**2+25*p[1]**2*p[2]-81*p[3]*
p[2]**2-225*p[2]**2*p[1]+18*
p[3]**2*p[2]+152*p[3]**2*p[1])/
p[2]/(9*p[3]*p[2]+16*p[3]*p[1]+
25*p[2]*p[1])    ;

    EQ_Z3.. Z("C3") =E=
1/9*(-236*p[3]*p[1]*p[2]+128*p[3]*
p[1]**2+875*p[1]**2*p[2]+108*p[3]*
p[2]**2+1200*p[2]**2*p[1]-585*
p[3]**2*p[2]-1040*p[3]**2*p[1])/
p[3]/(9*p[3]*p[2]+16*p[3]*p[1]+
25*p[2]*p[1])    ;
```

This "semi-GAMS" file is however is not ready to be processed by GAMS as it needs at least two adjustments to reconcile the following syntactical differences between GAMS and Maple.

- Firstly, Maple uses square brackets [ ] for array elements while GAMS uses combinations of parentheses and double quotes (" "). This can be adjusted by either manually editing or using Unix utilities such as

```
sed -e 's/\[/("C/g'
    -e 's/\]/")/g'
gexmaple.eqn > gexmaple.inc
```

- Secondly, the `lprint` routine of Maple is not sophisticated enough to be able to break lines at relevant breaking points. Instead, it simply folds lines at length specified by `screenwidth` (default 79 characters). If the line is split right in the middle of two double quotes delimiting an array element as in the following example:

```
-1/9*(-170*p("C3")*p("C1   <-----
")*p("C2")+656*p("C3        <-----
")*p("C1")**2
```

GAMS will issue a compiler error message. This can be fixed by simply manually editing to remove those "bad" newline characters and join the lines together as follows:

```
-1/9*(-170*p("C3")*p("C1")*
p("C2")+656*p("C3")*
p("C1")**2
```

These idiosyncratic differences can be interpreted as the weakness of either Maple or GAMS (or both) depending on one's inclination to favor one software package over the other. Our intention here is not to point a finger at whom is to be blamed, but rather to show that these problems surface only if modelers attempt to integrate both tools in the modeling process. It is probably not too technically difficult for developers of these software packages to provide a conversion routine between Maple and GAMS outputs.

Listing of the output file `gexmaple.eqn` after manual editing and renaming as `gexmaple.inc` is as follows:

```
*****  GAMS FILE    gexmaple.inc  *****
*****  originally   gexmaple.eqn  *****
*****  included by  gexmaple.gms  *****
EQUATIONS
    EQ_Z1   market excess demand
    EQ_Z2   market excess demand
    EQ_Z3   market excess demand
    EQ_object  objective function;

    EQ_object.. object =E=
sum(j,(p(j) * Z(j)) ** 2);
```

```
   EQ_Z1.. Z("C1") =E=
-1/9*(-170*p("C3")*p("C1")*p("C2")+
656*p("C3")*p("C1")**2+1025*
p("C1")**2*p("C2")-378*p("C3")*
p("C2")**2-150*p("C2")**2*
p("C1")-477*p("C3")**2*p("C2")-
128*p("C3")**2*p("C1"))/p("C1")/
(9*p("C3")*p("C2")+16*p("C3")*
p("C1")+25*p("C2")*p("C1"))    ;

   EQ_Z2.. Z("C2") =E=
2/3*(11*p("C3")*p("C1")*p("C2")+
88*p("C3")*p("C1")**2+25*
p("C1")**2*p("C2")-81*p("C3")*
p("C2")**2-225*p("C2")**2*
p("C1")+18*p("C3")**2*p("C2")+
152*p("C3")**2*p("C1"))/p("C2")/
(9*p("C3")*p("C2")+16*p("C3")*
p("C1")+25*p("C2")*p("C1"))    ;

   EQ_Z3.. Z("C3") =E=
1/9*(-236*p("C3")*p("C1")*p("C2")+
128*p("C3")*p("C1")**2+875*
p("C1")**2*p("C2")+108*p("C3")*
p("C2")**2+1200*p("C2")**2*
p("C1")-585*p("C3")**2*p("C2")-
1040*p("C3")**2*p("C1"))/p("C3")/
(9*p("C3")*p("C2")+16*p("C3")*
p("C1")+25*p("C2")*p("C1"))    ;
```

4.5 The Maple-Assisted GAMS Solution

Output files gexparam.inc and gexmaple.inc from the Maple interface gexmaple.mpl now can be included in the following GAMS program gexmaple.gms which produces exactly the same general equilibrium solution as that reported by the manually-coded GAMS program gexgmas.gms in subsection 3.2.

```
*****  GAMS FILE  gexgams.gms  *****
*****  Maple-assisted         *****

$INCLUDE "gexparam.inc"

POSITIVE VARIABLES
   p(j)       prices;

FREE VARIABLES
   Z(j)       market excess demands
   object     objective function;
```

```
* ------------------------------------
* include equation block generated
* by the Maple interface gexmaple.mpl
$INCLUDE "gexmaple.inc"
* ------------------------------------

MODEL gex /ALL/;

p.LO(j) = 0.000001;
p.L(j)  = 1.0;
Z.L(j)  = 0;

SOLVE gex MINIMIZING object USING NLP;
```

There are some structural differences between the *Maple-assisted* GAMS program in this section and the equation block of the *manually coded* GAMS program gexgams.gms of subsection 3.2. The manually coded GAMS program follows the algebraic structure of the model as closely as possible by writing equations with general index sets j, i as abbreviations of a multitude of equations numerically indexed by $j = 1$, $j = 2$, $j = 3$ and $i = 1$, $i = 2$. On the other hand, the Maple-assisted program writes equations as final expressions with the indices even being evaluated to numerical values $1, 2, 3$ for j and $1, 2$ for i. That is, the elegance of GAMS as an "general algebraic modeling system" language was sacrificed for the practical symbolic manipulation capability of Maple. It is probably still a long way for Maple to develop an ability which can, like the human brain, extract and reduce unwieldy arithmetic details to simple and meaningful intelligent expressions, i.e., *"artificial theorizing"*.

The following table provides a comparison of statistics on GAMS solutions by the manually coded program gexgams.gms and the Maple assisted program gexmaple.gms both executed on a 33Mhz 486 personal computer with 8 Mbytes of memory running DOS 5.0. As expected, because the Maple assisted program is unable to write compact formulas for the equation block, it is more bulky with code length almost double (1.9004 ratio) that of the manually coded counterpart. This explains a thirty percent (1.29 ratio) longer time required for compilation and execution. However, its model specification has two-thirds fewer equations (0.3333 ratio) and one half fewer variables (0.4667 ratio) which results in one fourth smaller resource usage (0.7562 ratio) and less than one half as many iterations and objective function calls for the MINOS solver.

	Manually Coded (1)	Maple Assisted (2)	Ratios $\frac{(2)}{(1)}$
Model Statistics			
Single equations	12	4	0.3333
Single variables	15	7	0.4667
Nonzero elements	48	19	0.3958
Derivative pool	9	11	1.2222
Code length	522	992	1.9004
Constant pool	11	25	2.2727
Time in Seconds			
Compilation	0.170	0.220	1.2941
Execution	0.380	0.490	1.2895
Total	0.550	0.710	1.2909
Solver Statistics			
Resource usage	1.813	1.371	0.7562
Iteration	42	23	0.5476
Obj. func. calls	104	49	0.4712

These crude statistics however did not reveal the hidden costs associated with time resource and effort spent by modelers who, if without the help of Maple, would have to perform the tedious and error-prone manual process of analytical derivation of the equation block required by GAMS. Measurements of these costs remain to be estimated but we would not be surprised if they could be substantial.

5 Conclusion

Symbolic computation software such as Maple can indeed be useful to CGE modelers in the model-building process. An interface between Maple and specialized modeling software such as GAMS provides the best of both worlds: mathematical manipulation from Maple and large-scale numerical computation from GAMS. The paper presents a simple illustrative example showing how Maple can be successfully used in conjunction with GAMS.

A combination of two powerful tools such as GAMS and Maple now allows CGE modelers to explore areas of research which used to be considered technically forbidden. For example, with Maple's ability to derive explicit functional forms for market excess demands and GAMS's ability to numerically solve for general equilibrium price solutions, stability and uniqueness conditions for general equilibrium solutions can be directly checked either analytically or numerically or both. Consequently modelers will not have to resort to an inefficient grid search (Kehoe and Whalley [1985]) nor simply assume away the problem as often practiced in the field.

References

[1] Brooke, A., D. Kendrick, and A. Meeraus (1992) *GAMS 2.25 A User's Manual.* San Francisco: The Scientific Press.

[2] Char, B., et al. (1991a) *Maple V Language Reference Manual.* New York: Springer-Verlag.

[3] Char, B., et al. (1991b) *Maple V Library Reference Manual.* New York: Springer-Verlag.

[4] Debreu, G. (1959) *Theory of Value.* New York: Wiley.

[5] Kehoe, T. J., and J. Whalley (1985) "Uniqueness of Equilibrium in Large Scale Numerical General Equilibrium Models," *Journal of Public Economics* 28, 247–254.

[6] Mansur, A., and J. Whalley (1984) "Numerical Specification of Applied General Equilibrium Models: Estimation, Calibration, and Data," chapter 3 in Scarf, H. E. and J. B. Shoven (eds.), *Applying General Equilibrium.* New York: Cambridge University Press.

[7] Meeraus, A. (1983) "An Algebraic Approach to Modeling," *Journal of Economic Dynamics and Control* 5:1, 81–108.

[8] Merrill, O. H. (1972) "Applications and Extension of an Algorithm That Computes Fixed Points of Certain Upper Semi-Continuous Point to Set Mappings," Ph. D. thesis, Department of Industrial Engineering, University of Michigan.

[9] Murtagh, B. A. and M. A. Saunders (1987) "MINOS 5.1 User's Guide," Report SOL 83-20R, December 1983. revised January 1987, Stanford University.

[10] Scarf, H. E. (with T. Hansen) (1973) *The Computation of Economic Equilibria*. New Haven: Yale University Press.

[11] Scarf, H. E. and J. B. Shoven, eds. (1984) *Applied General Equilibrium Analysis*. New York: Cambridge University Press.

[12] Shoven, J. B., and J. Whalley (1992) *Applying General Equilibrium*. New York: Cambridge University Press.

[13] van der Laan, G., and A. J. J. Talman (1979) "A Restart Algorithm for Computing Fixed Points Without an Extra Dimension," *Mathematical Programming* 17, 74–84.

[14] Weintraub, E. R. (1983) "The Existence of a Competitive Equilibrium: 1930–1954," *Journal of Economic Literature* 21, 1–39.

Trien T. Nguyen is an associate professor of economics with publications and research interests in computable general equilibrium modeling of economic policy in public finance, international trade, and economic development. Contact address: Professor Trien T. Nguyen, Department of Economics, University of Waterloo, 200 University Avenue West, Waterloo, Ontario, Canada N2L 3G1. Telephone: (519) 885-1211 ext. 2794. Fax: (519) 725-0530. Email: nguyen@watserv1.uwaterloo.ca.

CALCULATION OF THE STATE TRANSITION MATRIX FOR LINEAR TIME VARYING SYSTEMS

J. Watkins, S. Yurkovich
Department of Electrical Engineering, Ohio State University, Columbus OH, USA

Abstract *

This paper discusses how Maple can be used to calculate the state transition matrix for several classes of linear time-varying (LTV) systems. The state transition matrix is essential in determining the complete solution, stability, controllability, and observability of LTV systems. Unfortunately, a closed form solution for the state transition matrix exists only when the LTV system satisfies certain properties. Maple routines are discussed which have been written to find the state matrix when these properties are satisfied. Examples and applications to control system design are given.

I. Introduction

Applications of linear, time-varying (LTV) systems include rocket dynamics, time-varying linear circuits, satellite systems, and pneumatic actuators. LTV structure is also often assumed in adaptive and standard gain-scheduled control systems. The motivation for this work is an investigation of an alternate approach to gain-scheduled control systems.

In particular, we are interested in LTV systems of the form

$$\begin{aligned} \dot{x}(t) &= A(t)x(t) + B(t)u(t), \\ y(t) &= C(t)x(t) + D(t)u(t), \end{aligned} \qquad (1)$$

where $x(t) \in \Re^n$ is the state vector, $u(t) \in \Re^m$ is the control input, and $y(t) \in \Re^p$ is the system output. The state transition matrix is the unique solution to

$$\dot{\Phi}(t, t_0) = A(t)\Phi(t, t_0), \quad \Phi(t_0, t_0) = I, \qquad (2)$$

where I is the identity matrix. The state transition matrix is essential in determining the complete solution, stability, controllability, and observability of (1). It is also useful in design of controllers and observers for (1).

Unfortunately, a closed form solution to (2) exists only when $A(t)$ satisfies certain properties. Maple can easily find the solution of (2) when $A(t)$ is constant or satisfies a well known commutative property. This paper describes routines written to expand the class of systems for which Maple can easily calculate $\Phi(t, t_0)$ to include systems where $A(t)$ is triangular or $A(t)$ satisfies certain bracket properties. Routines have also been written to calculate the general solution of (2) to any arbitrary order, to check when $A(t)$ satisfies the commutative property, and to check whether a given $\Phi(t, t_0)$ satisfies (2).

II. State Transition Matrix Properties

The state transition matrix is an integral component in the study of LTV systems of the form given by (1). It is used for determining the complete solution, stability, controllability, and observability. It can also be used in the design of controllers and observers for (1). In this section we will discuss these uses along with some of the properties of the state transition matrix.

The state transition matrix, which satisfies

$$\dot{\Phi}(t, t_0) = A(t)\Phi(t, t_0), \qquad (3)$$

and has the following important properties [1]:

$$\Phi(t, t) = I \qquad (4)$$

Mathematical Computation with Maple V:
Ideas and Applications
Tom Lee, Editor
©1993 Birkhäuser Boston

*This work was supported under the USAF Graduate Fellowship Program

$$\Phi^{-1}(t, t_0) = \Phi(t_0, t) \qquad (5)$$
$$\Phi(t_2, t_0) = \Phi(t_2, t_1)\Phi(t_1, t_0). \qquad (6)$$

Stability of the homogeneous system,

$$\dot{x}(t) = A(t)x(t), \qquad (7)$$

whose solution is given by

$$x(t) = \Phi(t, t_0)x_0, \qquad (8)$$

where $x_0 = x(t_0)$, can be determined from the state transition matrix, according to well known stability theorems [1] (see Appendix). The necessary and sufficient conditions on $\Phi(t, t_0)$ for stability are summarized in Table 1.

It is easy to verify that the solution to the non-homogeneous system (1) is given by

$$
\begin{aligned}
x(t) &= \Phi(t, t_0)x_0 + \int_{t_0}^{t} \Phi(t, \tau)B(\tau)u(\tau)d\tau \\
y(t) &= C(t)\Phi(t, t_0)x_0 + C(t) \qquad (9) \\
&\quad \int_{t_0}^{t} \Phi(t, \tau)B(\tau)u(\tau)d\tau + D(t)u(t).
\end{aligned}
$$

To guarantee that the system can be driven from one state x_0 to another state x_1 with an input $u(t)$, it is necessary to show that the system is *controllable*. The LTV system (1) is said to be *controllable* if given any x_0 there exists an input $u(t)_{[t_0, t_1]}$ such that $x(t_1) = 0$. Controllability of (1) can be determined from the state transition matrix according to a well known theorem [2] (see Appendix).

To guarantee that the system states x(t) can be estimated from the system output y(t), it it necessary to show that the system is observable. The LTV system (1) is said to be *observable* on $[t_0, t_1]$ if the initial state x_0 is uniquely determined by the output y(t) for $t \in [t_0, t_1]$. Observability of (1) can be determined from the state transition matrix according to a well known theorem [2] (see Appendix). The controllability and observability Gramians, $W(t_0, t_1)$ and $M(t_0, t_1)$, respectively, (see Appendix) can also be used in the design of controllers and observers for (1).

It is clear that the state transition matrix is important for studying stability, controllability, and observability of (1). Calculation of the state transition matrix for linear, time-invariant systems is a straight forward task. Unfortunately, for linear time varying systems, it is often difficult if not impossible to calculate the state transition matrix.

III. Calculating the State Transition Matrix

In general, a closed form solution for $\Phi(t, t_0)$ does not exist. In this section, several classes of systems for which $\Phi(t, t_0)$ can be calculated in closed form are discussed. Maple routines which aid in this calculation are discussed. After summarizing classes for which the state transition can be calculated, two decomposition schemes will be discussed which expand on these classes.

Before examining these classes, it should be noted that the procedures to be discussed assume that the **linalg** package has been loaded along with the procedures listed in Table 2. The **eye** and **zeros** procedures are modeled after the Matlab functions of the same names and return the identity matrix and a matrix of zeros, respectively. The **msubs** procedure substitutes expressions into matrices.

The Kronecker product of $A = [a_{ij}] \in \Re^{m \times n}$ and $B \in \Re^{p \times q}$ is defined to be

$$
A \otimes B \equiv
\begin{bmatrix}
a_{11}B & \cdots & a_{1n}B \\
\vdots & \ddots & \vdots \\
a_{m1}B & \cdots & a_{mn}
\end{bmatrix}
\qquad (10)
$$

where $A \otimes B \in \Re^{mp \times qn}$. The vector vec A is defined to be

$$
\text{vec } A \equiv
\begin{bmatrix}
\bar{a}_1 \\
\vdots \\
\bar{a}_n
\end{bmatrix},
\qquad (11)
$$

where $A \in \Re^{m \times n}$, vec $A \in \Re^{mn}$, and the $\bar{a}_i$ are the columns of the matrix A. The command **invec(vec(A),m,n)** will return the matrix A. The procedures **kron** and **vec** transform matrix equations of the form

$$AXB = C \qquad (12)$$

where $A \in \Re^{m \times n}$, $X \in \Re^{n \times p}$, and $B \in \Re^{p \times q}$ and X is unknown to the equivalent system of equations

$$(B^T \otimes A)\text{vec } X = \text{vec } C. \qquad (13)$$

This equation can be solved for vec X by **linsolve(D, vec C)** where $D = B^T \otimes A$. [3]

With these tools, we can now study the solution of

$$\dot{\Phi}(t, t_0) = A(t)\Phi(t, t_0), \quad \Phi(t_0, t_0) = I. \qquad (14)$$

While in general there is no closed form solution to (14), the solution can be expressed in terms of

Table 1: Stability Bounds on $\Phi(t, t_0)$

Stable in the sense of Lyapunov at t_0	$\|\Phi(t,t_0)\| < k(t_0) < \infty$
Uniformly stable in the sense of Lyapunov	$\|\Phi(t,t_0)\| < k < \infty$
Asymptotically stable at t_0	$\|\Phi(t,t_0)\| \le k(t_0) < \infty$
	$\|\Phi(t,t_0)\| \to 0$ as $t \to \infty$
Uniformly asymptotically stable	$\|\Phi(t,t_0)\| \le k_1 e^{-k_2(t-t_0)}$

Table 2: Maple Utility Procedures

eye($\cdot$)	Computes identity matrix
zeros($\cdot$)	Computes matrix of zeros
msubs($A, s_1, ..., s_n$)	Substitutes subexpressions into matrix
kron(A,B)	Kronecker product $A \otimes B$
vec(A)	vec A
invec(A,m,n)	inverse of vec A

the Peano-Baker series [2]

$$\Phi(t,t_0) = I + \int_{t_0}^{t} A(\tau_1)d\tau_1 \qquad (15)$$
$$+ \int_{t_0}^{t} A(\tau_1) \int_{t_0}^{\tau_1} A(\tau_2)d\tau_2 d\tau_1 + \cdots.$$

The following procedure, **peano(A,n,t0)**, allows us to calculate this series to any desired order:

```
peano:=
proc(A,n,t0)
local w1,w2,i,tp;
 if not type(n,posint) then
  ERROR('n must be positive integer')
 elif not type(A,matrix) then
  ERROR('A must be matrix')
 elif not type(t0,scalar) then
  ERROR('t0 must be a scalar')
 fi;
 w2 := eye(A);
 w1 := evalm(w2);
 for i to n-1 do
  w2:=map(int,multiply(A,w2),t=t0..tp);
  w2:=msubs(msubs(w2,t = t.i),tp = t);
  w1:=add(w1,w2)
 od;
 RETURN("")
end
```

For time invariant systems, that is, when $A(t) = A$ where A is a *constant* matrix, the state transition matrix is given as

$$\Phi(t,t_0) = e^{A(t-t_0)}. \qquad (16)$$

When A is time invariant, Maple can easily calculate $\Phi(t,t_0)$ using **exponential(A, t-t0)**.

If $A(t)$ satisfies the *commutative* property,

$$A(t)\left(\int_{t_0}^{t} A(t_1)dt_1\right) = \left(\int_{t_0}^{t} A(t_1)dt_1\right)A(t), \qquad (17)$$

for all t, t_0, the state transition matrix is given as

$$\Phi(t,t_0) = e^{\int_{t_0}^{t} A(t_1)dt_1}. \qquad (18)$$

Again, Maple can easily calculate $\Phi(t,t_0)$ using the command **exponential(map(int, msubs(A, t=t1), t1=t0..t))**.

Checking (17) is equivalent to checking if

$$A(t_1)A(t_2) = A(t_2)A(t_1), \ \forall t_1, t_2 \qquad (19)$$

is satisfied [4]. A procedure **com** has been written to check condition (19). The commutative condition can also be checked by decomposing $A(t)$ into the form

$$A(t) = \sum_{i=1}^{q} \alpha_i(t)A_i \qquad (20)$$

159

where the $\alpha_i(t)$'s are linearly independent time functions and the A_i's are constant matrices, and then checking if the A_i's form a commuting family, that is $A_i A_j = A_j A_i$, $\forall i, j$. When the system is commutative, the state transition matrix is given by

$$\Phi(t, t_0) = \prod_{i=1}^{q} e^{A_i \beta_i(t, t_0)}, \qquad (21)$$

where $\beta_i(t, t_0) = \int_{t_0}^{t} \alpha_i(\tau) d\tau$ [5]. The solution of (19) and (21) are equivalent; however, the solution of (21) may result in a simpler form. For another decomposition method, see [6]. Clearly, constant matrices and diagonal matrices both satisfy the commutative property.

If $\dot{A}(t)$ exists and there is a constant matrix A_1 which satisfies

$$\frac{d}{dt} A(t) = A_1 A(t) - A(t) A_1 \ \forall t \geq t_0, \qquad (22)$$

then

$$\Phi(t, t_0) = e^{A_1(t-t_0)} e^{A_2(t-t_0)} \ \forall t \geq t_0, \qquad (23)$$

where $A_2 = A(t_0) - A_1$ [7]. If $A(t)$ satisfies (22), it is said to be a member of the $\mathcal{A}_1$ class. A necessary condition to satisfy (22) is that the eigenvalues of $A(t)$ are time invariant. Stability for members of $\mathcal{A}_1$ can be completely determined from the eigenvalues of A_1 and A_2.

The Kronecker product is used to transform (22) to

$$(A^T \otimes I - I \otimes A) \text{vec } A_1 = \text{vec } \dot{A}, \qquad (24)$$

which can be solved for vec A_1 using **linsolve**. The following procedure, **fa1(A,A0,A1,A2)**, transforms (22) to (24) and returns A_0, A_1, and A_2:

```
fa1:=
proc(A,A0,A1,A2)
local ad,Ad,a1,Al,n;
 n := linalg[rowdim](A);
 Ad := map(diff,A,t);
 ad := vec(Ad);
 Al := add(kron(linalg[transpose](A),
       eye(A)),-kron(eye(A),A));
 a1 := linalg[linsolve](Al,ad);
 if a1 = NULL then
  ERROR('no solution possible') fi;
 A0 := msubs(A,t = 0);
 A1 := invec(a1,n,n);
 A2 := evalm(A0-A1)
end
```

We note that multiple solutions are possible; the user should therefore choose the free parameters such that A_1 and A_2 are time invariant if possible. If a time invariant solution exists, $\Phi(t, t_0)$ is given by **multiply(exponential(A1, t-t0), exponential(A2, t-t0))**.

The class $\mathcal{A}_1$ can be extended as follows [8]. If $\dot{A}(t)$ and $\dot{h}(t)$ exist, where $h(t)$ is a nonzero scalar time function, and there is a constant matrix A_1 which satisfies

$$\frac{d}{dt} \left(\frac{A(t)}{h(t)} \right) = A_1 A(t) - A(t) A_1 \ \forall t \geq t_0, \qquad (25)$$

then

$$\Phi(t, t_0) = e^{A_1 g(t)} e^{A_2 g(t)} \ \forall t \geq t_0, \qquad (26)$$

where $g(t) = \int_{t_0}^{t} h(\tau) d\tau$ and $A_2 = \lim_{t \to t_0} \left(\frac{A(t)}{h(t)} \right) - A_1$. If $A(t)$ satisfies (25), it is said to be a member of the $\mathcal{A}_h$ class. A necessary condition to satisfy (25) is that the eigenvalues of $A(t)$ are constants multiplied by $h(t)$. This can be checked using **eigenvals(A)**. When $h(t) = 1$ this class reverts to the $\mathcal{A}_1$ class, and when $h(t) = 1/t$ the Euler-type differential equation can be solved [9].

The following procedure, **fah(A,Ah0,A1,A2,t0,h,g)**, also uses the Kronecker product and returns A_1, A_2 and g:

```
fah:=
proc(A,Ah0,A1,A2,t0,h,g)
local ad,Ad,a1,Al,n,Ah,t1;
 g := eval(int(subs(t = t1,h),t1=t0..t));
 n := linalg[rowdim](A);
 Ah := evalm(1/h*A);
 Ad := map(diff,Ah,t);
 ad := vec(Ad);
 Al := add(kron(linalg[transpose](A),
       eye(A)),-kron(eye(A),A));
 a1 := linalg[linsolve](Al,ad);
 if a1 = NULL then
  ERROR('no solution possible') fi;
 Ah0 := map(limit,Ah,t = t0);
 A1 := invec(a1,n,n);
 A2 := evalm(Ah0-A1)
end
```

Again, if a time invariant solution for A_1 and A_2 exists, $\Phi(t, t_0)$ is given by **multiply(exponential(A1,g),exponential(A2,g))**.

The state transition matrix can be found when $A(t)$ is *triangular*. If $A(t) = [a_{ij}(t)]$ is lower triangular, that is $a_{ij}(t) = 0$, $\forall j > i$, then the state transition matrix has elements given by

$$
\phi_{ij}(t,t_0) = \begin{cases} 0 & i < j \\ e^{\int_{t_0}^{t} a_{ii}(\tau)d\tau} & i = j \\ \int_{t_0}^{t} \phi_{ii}(t,\tau) \times \\ (\sum_{k=j}^{i-1} a_{ik}(\tau)\phi_{kj}(\tau,t_0))d\tau & i > j, \end{cases}
$$

where $\Phi(t,t_0) = [\phi_{ij}(t,t_0)]$. The following procedure, **ltriangle(A)**, calculates $\Phi(t,t_0)$ for lower triangular matrices:

```
ltriangle:=
proc(A)
local phi,n,i,sumk,k,j,l;
 n := rowdim(A);
 phi := map(0,A);
 for i to n do
  phi[i,i] := exp(int(subs(t = t1,
           A[i,i]),t1 = t0 .. t));
  for j to i-1 do
   l := i-j+1;
   sumk := subs(t = t.l,sum(A[i,k]*
          phi[k,j],k = j .. i-1));
   phi[i,j] := phi[i,i]*int(subs(t = t0,
            subs(t0 = t.l,phi[i,i]))*
            sumk,t.l = t0..t)
  od
od;
map(simplify,phi);
RETURN("")
end
```

If $A(t)$ is upper triangular, that is $a_{ij}(t) = 0$, $\forall i > j$, the state transition has elements given by

$$
\phi_{ij}(t,t_0) = \begin{cases} 0 & i > j \\ e^{\int_{t_0}^{t} a_{ii}(\tau)d\tau} & i = j \\ \int_{t_0}^{t} \phi_{ii}(t,\tau) \times \\ (\sum_{k=i+1}^{j} a_{ik}(\tau)\phi_{kj}(\tau,t_0))d\tau & i < j. \end{cases}
$$

Note that a *diagonal* matrix is both lower and upper triangular. The following procedure, **utriangle(A)**, calculates $\Phi(t,t_0)$ for upper triangular matrices:

```
utriangle:=
proc(A)
local phi,n,i,sumk,k,j,l;
 n := rowdim(A);
 phi := map(0,A);
 for i from n by -1 to 1 do
```

```
  phi[i,i] := exp(int(subs(t = t1,
            A[i,i]),t1 = t0 .. t));
  for j from i+1 to n do
   l := j-i+1;
   sumk := subs(t = t.l,sum(A[i,k]*
          phi[k,j],k = i+1 .. j));
   phi[i,j] := phi[i,i]*int(subs(t = t0,
            subs(t0 = t.l,phi[i,i]))*
            sumk,t.l = t0..t)
  od
od;
map(simplify,phi);
RETURN("")
end
```

Triangle matrices can be extended to *block triangular matrices*. Let $A(t) = [A_{ij}(t)]$, where each $A_{ij}(t)$ has dimensions $n_i \times n_j$, and $\Phi(t,t_0) = [\Phi_{ij}(t,t_0)]$, where $\Phi_{ij}(t,t_0)$ has dimensions $n_i \times n_j$. If $A(t)$ is lower block triangular, that is $A_{ij}(t) = 0$, $\forall j > i$, then the state transition has blocks given by

$$
\Phi_{ij}(t,t_0) = \begin{cases} 0 & i < j \\ \Phi_{ii} & i = j \\ \int_{t_0}^{t} \Phi_{ii}(t,\tau) \times \\ (\sum_{k=j}^{i-1} A_{ik}(\tau)\Phi_{kj}(\tau,t_0))d\tau & i > j. \end{cases}
$$
(27)

Likewise, if $A(t)$ is upper block triangular, that is $A_{ij}(t) = 0$, $\forall i > j$, then the state transition has blocks given by

$$
\Phi_{ij}(t,t_0) = \begin{cases} 0 & i > j \\ \Phi_{ii}(t,t_0) & i = j \\ \int_{t_0}^{t} \Phi_{ii}(t,\tau) \times \\ (\sum_{k=i+1}^{j} A_{ik}(\tau)\Phi_{kj}(\tau,t_0))d\tau & i < j. \end{cases}
$$
(28)

While (27) and (28) hold for all lower and upper block triangular matrices, respectively, $\Phi(t,t_0)$ can be calculated explicitly only when the $\Phi_{ii}(t,t_0)$'s are known. Therefore, if every $A_{ii}(t)$ is from a class of matrices from which $\Phi_{ii}(t,t_0)$ can be calculated, $\Phi(t,t_0)$ can also be calculated. Note that a *block diagonal* matrix is both lower and upper block triangular. While no Maple procedures have been written to work directly with block triangular matrices, the Maple procedures discussed earlier would make this calculation easier.

Two decomposition schemes exist which expand the class of systems for which $\Phi(t,t_0)$ can be calculated. The first one can be found in [10]. Let

$A(t)$ be decomposed as

$$A(t) = \sum_{i=1}^{m} A_i(t) \qquad (29)$$

such that

$$\dot{\Phi}_i(t,0) = F_i(t)\Phi_i(t,0), \quad i = 1, 2, \ldots, m \qquad (30)$$
$$\Phi_i(0,0) = I,$$

is solvable where

$$F_i(t) = T_{i-1}^{-1}(t)A_i(t)T_{i-1}(t) \quad i = 1, 2, \ldots, m,$$

and

$$T_i(t) = \Phi_i(t,0)T_{i-1}(t) \quad i = 1, 2, \ldots, m-1,$$
$$T_0(t) = I.$$

Then we have that

$$\Phi(t,0) = \prod_{i=1}^{m} \Phi_i(t,0). \qquad (31)$$

Again, while no Maple procedure has been written to deal with this decomposition scheme directly, the procedures discussed to this point are useful in doing so. It is interesting to note that this decomposition scheme can be used to show that any arbitrary $A(t)$ can be decomposed into two normal systems [11]. Consequently, if closed form solutions exist for these normal systems, a closed form solution exists for any arbitrary $A(t)$.

A second decomposition scheme can be found in [12] and [13]. Differentiating (7) results in

$$\ddot{x} = (\dot{A}(t) + A^2(t))x(t). \qquad (32)$$

If

$$\dot{A}(t) + A^2(t) = B(t)A(t), \qquad (33)$$

for some $B(t)$, then (32) can be rewritten as

$$\ddot{x} = B(t)\dot{x}. \qquad (34)$$

If a solution for $B(t)$ exists, the following procedure, **hemami(A)**, returns $B(t)$:

```
hemami:=
proc(A)
local ad,Ad,b,B,n,A1;
 n := linalg[rowdim](A);
 Ad := map(diff,A,t);
 ad := vec(evalm(Ad+A^2));
 A1 := kron(linalg[transpose](A),eye(A));
 b := linalg[linsolve](A1,ad);
 if b = NULL then
```

```
   ERROR('no solution possible') fi;
 B := invec(b,n,n)
end
```

If

$$\dot{\Phi}_1(t,0) = B(t)(t)\Phi_1(t,0), \quad \Phi_1(0,0) = I \qquad (35)$$

is solvable then

$$\Phi(t,t_0) = \left(\int_{t_0}^{t} \Phi_1(\tau,t_0)d\tau \right) A(t_0) + I. \qquad (36)$$

Assuming **Phi1** has been found, $\Phi(t,t_0)$ is given by **add(multiply(map(int, msubs (Phi1, t=t1), t1=t0..t), msubs(A, t=t0)), eye(A))**. If (35) is not solvable using any of the previous procedures, (34) can be differentiated. If a solution to the Riccati equation $\dot{B}(t) + B^2(t) = C(t)B(t)$ exists, this procedure can be repeated until $\Phi(t,t_0)$ is found, or until the Riccati equation has no solution.

Lastly, we discuss the procedure used to verify that a given $\Phi(t,t_0)$ satisfies (14). From (14), we know that

$$A(t) = \Phi^{-1}(t,t_0)\dot{\Phi}(t,t_0) \qquad (37)$$
$$\Phi(t_0,t_0) = I.$$

The following procedure **check(Phi)** returns $\Phi^{-1}(t,t_0)\dot{\Phi}(t,t_0)$ and $\Phi(t_0,t_0)$:

```
check:=
proc(phi,t,t0)
local a,Id;
 a := multiply(map(diff,phi,t),
      inverse(phi));
 Id := map(limit,phi,t = t0);
 RETURN(map(simplify,Id),map(simplify,a))
end
```

In this section, several classes for which the state transition matrix can be calculated were given. These include constant matrices, matrices which satisfy the commutative condition, the $\mathcal{A}_1$ class, the $\mathcal{A}_h$ class, triangular matrices, and block triangular matrices whose block diagonal matrices are solvable. Two decomposition schemes which extend these classes were also given. Maple procedures were given which aid in this calculation.

IV. Examples

In this section, we will apply the Maple procedures discussed earlier to two LTV systems. In exposition of the examples, we include diary-like output from the Maple implementation to assist the reader in reproducing these results.

Example 1

Consider Euler's equations for the rigid body [2]:

$$\dot{\omega}_1(t) = \frac{I_2 - I_3}{I_1}\omega_2(t)\omega_3(t) + \frac{1}{I_1}u_1(t)$$

$$\dot{\omega}_2(t) = \frac{I_3 - I_1}{I_2}\omega_3(t)\omega_1(t) + \frac{1}{I_2}u_2(t) \quad (38)$$

$$\dot{\omega}_3(t) = \frac{I_1 - I_2}{I_1}\omega_1(t)\omega_2(t) + \frac{1}{I_3}u_3(t)$$

where the ω_i are the angular velocities, the I_i are the principal moments of inertia, and the u_i are the applied torques. If we let $I_1 = I_2 = I$ (symmetrical body) and linearize about the nominal trajectory $\hat{u}_1 = \hat{u}_2 = \hat{u}_3 = 0$ and

$$\hat{\omega}_1(t) = \sin(wgt),$$
$$\hat{\omega}_2(t) = \cos(wgt),$$
$$\hat{\omega}_3(t) = w,$$

where $g = \frac{I - I_3}{I}$, we obtain the state equation

$$\dot{\bar{\omega}}(t) = A(t)\bar{\omega}(t) + B(t)u(t), \quad (39)$$

where $\bar{\omega}(t) = \omega(t) - \hat{\omega}(t)$ and

```
> A:=matrix([[0,g*w,g*cos(w*g*t)],
    [-g*w,0,-g*sin(w*g*t)],[0,0,0]]);

          [   0      g w     g cos(w g t)   ]
          [                                 ]
    A  := [ - g w     0     - g sin(w g t)  ]
          [                                 ]
          [   0       0           0         ]
```

The eigenvalues of $A(t)$, given by

```
> eigenvals(A);

          2  2 1/2            2  2 1/2
    0, (- g  w )     , - (- g  w )
```

are time invariant, which suggests that $A(t)$ may be in the $\mathcal{A}_1$ class. The procedure call **fa1(A,A0,A1,A2,0)** returns A_1 and A_2 with A_1 given below as

$$\begin{bmatrix} t_5 & t_4 & \frac{1}{w}(\cos(wgt)(t_4 - gw) + \sin(wgt)(t_5 - t_9)) \\ -t_4 & t_5 & \frac{1}{w}(\sin(wgt)(gw - t_4) + \cos(wgt)(t_5 - t_9)) \\ 0 & 0 & t_9 \end{bmatrix}$$

where t_4, t_5, and t_9 are parameters to be specified. These parameters are chosen such that A_1 and A_2 are time invariant,

```
> A1:=msubs(A1,t4=w*g,t5=0,t9=0);

              [   0     g w   0 ]
              [                 ]
        A1 := [ - g w    0    0 ]
              [                 ]
              [   0      0    0 ]

> A2:=msubs(A2,t4=w*g,t5=0,t9=0);

              [ 0   0   g ]
              [           ]
        A2 := [ 0   0   0 ]
              [           ]
              [ 0   0   0 ]
```

Now $\Phi(t,t_0)$ can be calculated as in (23) with

```
> phi:=map(simplify,multiply(
    exponential(A1,t),exponential(A2,t)));

          [cos(wgt)   sin(wgt)   cos(wgt) t g]
          [                                  ]
phi:=[-sin(wgt)  cos(wgt)  -sin(wgt) g t]
          [                                  ]
          [   0          0           1      ]
```

Finally, checking our results (according to (37))

```
> check(phi,t,0);

[1  0  0]  [  0      g w     g cos(w g t) ]
[       ]  [                              ]
[0  1  0], [- g w     0     - g sin(w g t)]
[       ]  [                              ]
[0  0  1]  [  0       0           0       ]
```

The identity matrix and the matrix $A(t)$ are returned as expected. Note that by examining $\Phi(t,0)$, we see that the nominal trajectory is unstable. The elements $\phi_{13}(t,0)$ and $\phi_{23}(t,0)$ of $\Phi(t,0)$ are clearly unbounded.

Example 2

Again consider Euler's equations for a symmetrical body, that is $I_1 = I_2 = I$. Setting $u_1 = u_2 = 0$, the equations can be written in the form of a quasi-linear parameter varying system [14]:

$$\frac{d}{dt}\begin{bmatrix} \omega_1(t) \\ \omega_2(t) \\ z(t) \end{bmatrix} = \begin{bmatrix} 0 & gz(t) & 0 \\ -gz(t) & 0 & 0 \\ 0 & 0 & 0 \end{bmatrix}\begin{bmatrix} \omega_1(t) \\ \omega_2(t) \\ z(t) \end{bmatrix}$$
$$+ \begin{bmatrix} 0 \\ 0 \\ \frac{1}{I_3} \end{bmatrix}u_3(t) \quad (40)$$

where $z(t) = \omega_3(t)$. A linear parameter varying (LPV) system depends on a time varying parameter rather than explicitly on time. This parameter

163

is known at time t, but not necessarily *a priori* [14].

The control structure $u_3(t) = -I_3k(z - z_{ref})$ is proposed to drive $z(t)$ to z_{ref} and maintain stability of the system where k is a positive constant. Substituting $u_3(t)$ into (40), we obtain the following closed loop matrix for $A(t)$,

```
> A:=matrix([[0,g*z(t),0],
    [-g*z(t),0,0],[0,0,-k]]);
```

```
      [   0      g z(t)    0  ]
      [                       ]
A :=  [ - g z(t)    0      0  ]
      [                       ]
      [   0         0    - k  ]
```

Taking the Peano-Baker series to the 5th order results in

```
> peano(A,5,0);
           2        4          3
  [1-1/2 %1  + 1/24 %1 , %2 - 1/6 %2 , 0]

          3             2       4
  [%1-1/6 %1 , 1 - 1/2 %2 + 1/24 %2 , 0]

                 2 2      3 3       4 4
  [0, 0, 1-k t+1/2 k t -1/6k t +1/24k t ]
```

```
              t
             /
            |
%1 :=       |  - g z(t1) dt1
            |
             /
             0

              t
             /
            |
%2 :=       |  g z(t1) dt1
            |
             /
             0
```

By examining the Peano-Baker series and checking a table of Taylor series, it is proposed that $\Phi(t,0)$ is of the following form:

```
> h:=int(g*z(t1),t1=0..t):
> phi:=matrix([[cos(h),sin(h),0],
    [-sin(h),cos(h),0],[0,0,exp(-k*t)]]);
```

```
      [ cos(%1)   sin(%1)       0      ]
      [                                ]
```

```
phi:=[ - sin(%1)  cos(%1)       0      ]
      [                                ]
      [    0          0     exp(- k t) ]
```

```
              t
             /
            |
%1 :=       |  g z(t1) dt1
            |
             /
             0
```

Checking our proposed $\Phi(t,0)$,

```
> check(phi,t,0);
```

```
[ 1  0  0 ] [    0      g z(t)    0  ]
[         ] [                        ]
[ 0  1  0 ],[ - g z(t)    0      0  ]
[         ] [                        ]
[ 0  0  1 ] [    0         0    - k  ]
```

the identity matrix and $A(t)$ are returned as expected. By examining $\Phi(t,0)$, it can be seen that ω_1 and ω_2 are stable and that z is asymptotically stable as desired.

V. Conclusion

In this paper, we have examined how Maple could be used to calculate the state transition matrix for several classes of LTV systems. We motivated the problem by arguing that the state transition matrix is important for understanding stability, controllability, and observability of LTV systems. Classes of systems for which the state transition matrix can be calculated include constant matrices, matrices which satisfy a commutative condition, the $\mathcal{A}_1$ class, the $\mathcal{A}_h$ class, triangular matrices, and block triangular matrices whose block diagonal matrices are solvable. Maple procedures to aid in this calculation were given. Finally, examples showing how the state transition matrix can be calculated and used for analysis of LTV systems were discussed.

Appendix

In the following we list several well known Theorems (involving the state transition matrix) which are referenced in the text.

1. Stability [1]:

Theorem 1 *Every equilibrium state of (7) is* stable in the sense of Lyapunov *at t_0 if and only if there exists some constant k which depends on t_0 such that $\|\Phi(t,t_0)\| \leq k < \infty$ for all $t \geq t_0$. If k is independent of t_0, it is* uniformly stable in the sense of Lyapunov.

2. Asymptotic Stability [1]:

Theorem 2 *The zero state of (7) is* asymptotically stable *at t_0 if and only if $\|\Phi(t,t_0)\| \leq k(t_0) < \infty$ and $\|\Phi(t,t_0)\| \to 0$ as $t \to \infty$. The zero state is* uniformly asymptotically stable *over $[0,\infty)$ if and only if there exist positive numbers k_1 and k_2 such that $\|\Phi(t,t_0)\| \leq k_1 e^{-k_2(t-t_0)}$ for any $t_0 \geq 0$ and for all $t \geq t_0$.*

3. Controllability [2]:

Theorem 3 *The LTV system (1) is* controllable *on $[t_0, t_1]$ if and only if the controllability Gramian*

$$W(t_0, t_1) = \int_{t_0}^{t_1} \Phi(t_0, t)B(t)B^T(t)\Phi^T(t_0, t)dt$$

is invertible.

4. Observability [2]:

Theorem 4 *The LTV system (1) is* observable *on $[t_0, t_1]$ if and only if the observability Gramian*

$$M(t_0, t_1) = \int_{t_0}^{t_1} \Phi^T(t_0, t)C^T(t)C(t)\Phi(t_0, t)dt$$

is invertible.

References

[1] C. Chen, *Linear System Theory and Design.* Orlando, Florida: Holt, Rinehart and Winston, Inc., 1984.

[2] W. J. Rugh, *Linear System Theory.* Englewood Cliffs, New Jersey: Prentice Hall, 1993.

[3] R. A. Horn and C. R. Johnson, *Topics in Matrix Analysis.* New York: Cambridge University Press, 1991.

[4] B. A. Kinariwala, "Analysis of time varying networks," in *IRE Int. Conv. Record, Part 4*, pp. 268–276, 1961.

[5] M. Wu and A. Sherif, "On the commutative class of linear time-varying systems," *International Journal of Control*, vol. 23, no. 3, pp. 433–444, 1976.

[6] L. A. Zadeh and C. A. Desoer, *Linear system theory; the state space approach.* New York: McGraw-Hill, 1963.

[7] M. Y. Wu, "A new method of computing the state transition matrix of linear time-varying systems," in *Proceedings of the IEEE International Symposium on Circuits and Systems*, pp. 269–272, San Francisco, CA, April 1974.

[8] M. Wu, "Solution of certain classes of linear time-varying systems," *International Journal of Control*, vol. 31, no. 1, pp. 11–20, 1980.

[9] E. A. Coddington and N. Levinson, *Theory of Ordinary Differential Equations.* New York: McGraw-Hill, 1955.

[10] M. Wu, "A successive decomposition method for the solution of linear time-varying systems," *International Journal of Control*, vol. 33, no. 1, pp. 181–186, 1981.

[11] J. Zhu and C. D. Johnson, "New results in the reduction of linear time-varying dynamical systems," *Siam Journal of Control and Optimization*, vol. 27, no. 3, pp. 476–494, May 1989.

[12] H. Hemami, "On a class of time-varying systems," *IEEE Transactions on Automatic Control*, vol. AC-13, no. 5, p. 589, October 1968.

[13] T. L. Niemeyer, "Exact state transition matrix solutions to time-varying systems," M.S. Thesis, The Ohio State University, 1969.

[14] J. S. Shamma and J. R. Cloutier, "Trajectory scheduled missile autopilot design," in *Conference on Control Applications*, pp. 237–242, Dayton, OH, September 1992.

John Watkins (watkins@ee.eng.ohio-state.edu) received the B.S. degree in electrical engineering from the University of Nebraska-Lincoln in 1989 and the M.S. degree in electrical engineering from The Ohio State University in 1991. He is currently pursuing his Ph.D. degree in electrical engineering at The Ohio State University where he is supported by the United States Air Force Laboratory Graduate Fellowship Program. His research interests include nonlinear and linear control systems, system identification, and the control of flexible mechanical structures.

Stephen Yurkovich (s.yurkovich@ieee.org) received the B.S. degree in engineering science from Rockhurst College in Kansas City, Missouri, in 1978, and the M.S. and Ph.D. degrees in electrical engineering from the University of Notre Dame in 1981 and 1984, respectively. He held teaching and postdoctoral research positions at Notre Dame in 1984 prior to moving later that year to the Department of Electrical Engineering at The Ohio State University, where he is currently Associate Professor. He has published extensively in the fields of system identification and control, and in application areas including flexible mechanical structures, robotic systems and automotive systems. Prof. Yurkovich teaches a wide range of undergraduate and graduate level courses in control theory, and has authored the text *Control Systems Laboratory*. He is past Chairman of the Standing Committee on Student Activities, and immediate past Chairman of the Standing Committee on Publications, in the IEEE Control Systems Society, where he is currently an Executive Officer and member of the Board of Governors. In January 1993, Dr. Yurkovich became Editor in Chief of IEEE *Control Systems*.

IV

MAPLE V IN SCIENCE AND ENGINEERING

Part B: Design

ALGEBRAIC COMPUTER AIDED-DESIGN WITH MAPLE V 2

C.T. Lim, M.T. Ensz, M.A. Ganter, D.W. Storti
Department of Mechanical Engineering, University of Washington, Seattle WA, USA

Introduction

This paper describes `implicit_solids` which implements implicit solid modeling (ISM) using MapleV2. Implicit solid modeling refers to the method that uses an implicit function to define a composite object. Various primitives are pre-defined and joined using Boolean operations to form the resulting implicit function [Storti et al. 1992].

A solid model contains an informationally complete description of the geometry and topology of a 3-D object [Requicha and Voelcker 1982]. There are many applications in which solid modeling can be used to adequately represent the geometry of a solid object before the object is manufactured. There are numerous advantages of constructing a model, notably models that have physical properties that are good approximations to those of the real object. Mass properties of the object such as its volume and moments of inertia can be computed. Design engineers can develop prototypes in a computer to visualize their design and determine the characteristics of the components. Using solid modeling not only allows visualization of objects without having to first manufacture them, it also permits changes to be made to the design with minimal effort.

There are three traditional categories of solid-modeling systems, namely, boundary representation (B-rep), spatial decomposition, and constructive solid geometry (CSG). The implementation of boundary representation involves the description of the enclosed surfaces, which can include planar polygons, quadrics, spline surfaces, and surface patches. Constraints on the type of surfaces used in the scheme affect the level of representation power. Furthermore, the order of the surface function polynomials has to be kept relatively low to maintain a reasonable level of complexity, which restricts the effectiveness of the scheme.

Spatial decomposition schemes include spatial enumeration, cell decomposition, and octree representation. Spatial enumeration represents an object as a collection of volume cells. Object representation is relatively easy to maintain but difficult to create due to the simplistic structure of the scheme. Shapes composed of large curvature variation are most difficult to model since they require more geometry than is provided by cubic cells. Cell decomposition is a generalization of spatial enumeration and allows the cells to vary in shape and size. Octree representation permits recursive subdivision of partially occupied cells, but requires significantly more memory as the size of the cells becomes smaller. In constructive solid geometry, objects are built by moving, joining, and cutting primitive objects. More specifically, primitive solids such as spheres, cuboids, and superellipsoids are combined using Boolean operations which include union, intersection, and difference. The resulting object is a tree-structured representation with the primitive solids at the leaves and the Boolean operations at the nodes. The scope of the representation is determined by the number and types of primitives allowed. Constructive solid geometry is attractive because it resembles manufacturing process and is easy to understand.

Implicit Solid Modeling

In implicit solid modeling (ISM), a solid object in the 3-D Euclidean space E^3 is described by a defining function of the spatial coordinates. The complete representation of the object is unambiguous and is associated with the ability to construct a point membership classification (PMC) which categorizes any point in E^3 as either belonging to the interior, exterior, or surface of the object. The defining function divides the space into three regions: the interior where $F(x,y,z) < f$, the exterior where $F(x,y,z) > f$, and the surface where $F(x,y,z) = f$ [Ricci 1973]. The constant f is referred to as the threshold value of the function.

Functional definitions of some useful primitives are readily available. For example, superellipsoids [Barr 1981] are defined by the function:

Mathematical Computation with Maple V:
Ideas and Applications
Tom Lee, Editor
©1993 Birkhäuser Boston

$$\left[\left(\frac{x}{a}\right)^{\frac{2}{e2}} + \left(\frac{y}{b}\right)^{\frac{2}{e2}} \right]^{\frac{e2}{e1}} + \left(\frac{z}{c}\right)^{\frac{2}{e1}} = 1 \qquad (1)$$

where a, b, and c define the geometric extent while e_1 and e_2 specify the shape properties. This function describes a broad family of easily defined primitives. Displacement and scaling of the primitives are achieved by linear coordinate transformations. Employing the basic primitives and operations, simple objects are represented with a selection comparable to existing CSG modelers, and additional primitives can be added by simply providing the appropriate defining functions.

While simple primitives are useful in forming a working basis for the modeling scheme, functional representations of more complicated objects produced by Boolean volume set operations are also necessary. Consider forming a function describing the intersecting region of the two solids defined by F_1 and F_2. Since all the points in the region have to satisfy both the inequalities $F_1 < f$ and $F_2 < f$, the intersection of two solids is associated with the defining function Maximum (F_1, F_2).

$$F_{1 \cap 2} = \text{Max}(F_1, F_2) \qquad (2)$$

In addition to intersection, other Boolean operations such as union and difference can be defined using Minimum and Inverse, in addition to Maximum.

To approximate the Maximum function with the l^n-norm (note that $\text{Max}(F_1, F_2) = l^{\infty}(|F_1|, |F_2|)$), Ricci [Ricci 1973] proposed a power law relation

$$F_{1 \cap 2} = (F_1 + F_2)^{1/n} \qquad (3)$$

The functions F_1 and F_2 are required to be non-negative and continuous in E^3. Staying within the algebraic realm allows the use of additional tools for variable elimination. Since the order of the blend, n, determines the accuracy of the approximation to the intersection, a larger n produces blending surfaces that cling more closely to the primitive solids. Note that blending effects are primarily noticeable near surface intersections. This approximation method automatically produces geometrically continuous blended surfaces. Although the discussion above pertains to two solids, the method can be applied simultaneously to any number of solids.

`Implicit_solids` uses Ricci's formulation with the threshold constant $f = 1$. The consequence of this choice is that the nth root in the power law relation can be deleted without altering the extent of the solid. For example, a sphere of radius 4 centered about the point $(0, 0, 0)$ is expressed as:

$$\text{sphere} = \left(\frac{x}{4}\right)^2 + \left(\frac{y}{4}\right)^2 + \left(\frac{z}{4}\right)^2 \qquad (4)$$

The resulting equations which give the modified-Ricci intersection, union, and difference of the algebraic-Boolean operators of degree n are:

$$F_{1 \cap 2} = F_1^n + F_2^n \qquad (5)$$

$$F_{1 \cup 2} = (F_1^{-n} + F_2^{-n})^{-1} \qquad (6)$$

$$F_{1-2} = F_1^n + F_2^{-n} \qquad (7)$$

Computer Algebra System

Even though the foundations of ISM have existed for 30 years, practical implementations and availability to the mass engineering audience were not possible. Before the advent of computer algebra systems, the usefulness of implicit solid modeling was limited because the defining functions for even moderately complicated solids became meaningless to the user. Now, this limitation can be overcome with the help of a computer algebra system such as MapleV2. Handling and manipulating functions that were once thought of as impossible are now not only feasible, but relatively effortless and error-free. The current implementation of MapleV2 on a workstation or PC provides sufficient computing power to accomplish basic ISM calculations.

In addition to the function evaluation and manipulation capabilities, MapleV2 also provides commands such as `implicitplot3d` and `implicitplot` to visualize solids and 2-D sections defined by implicit expressions. All of the figures in this paper were generated using either of the above mentioned commands except the ray-traced pictures in Figures 3 and 5. One minor limitation of the plotting commands is that the grid spacing has to be kept below a relatively low limit, which sometimes makes the plots appear "rough". An alternative around this problem is to perform a ray-trace on the defining function of the solid when a detailed picture of the solid is needed.

`Implicit_solids` offers the benefits of compact code, easy extension or modification, and a simple user interface, taking advantage of MapleV2's argument type checking and visualization capabilities. MapleV2's improvements in output display capabilities help ease the strain of examining long and complicated equations. The users are now able to choose various output display options, the most useful one being the high-resolution print which displays equations in near text-book form.

The proper usage of all the commands in `implicit_solids`, including the Boolean operators and coordinate transformations, can be found in the help pages. The help pages are included to provide the users with adequate information and guidance when needed and they were designed to conform to the standard MapleV2 help format. Some useful examples are also included in the help pages so that new users may start using the package by just duplicating the examples and observing the results.

Algebraic Solid Modeling

Computer algebra systems are most effective when dealing with algebraic problems, i.e. manipulating and solving polynomial equations. To adapt ISM to perform efficiently on computer algebra systems, minor restrictions are required, and the resulting scheme is called Algebraic Solid Modeling (ASM) [Ganter and Storti 1992]. In this scheme, solids are represented by implicit rational functions which are most suitable for use in computer algebra systems.

One of the constraints of ASM is that exponents used in defining superelliptic primitives and modified Ricci algebraic-Boolean operations must be integers, so that the resulting expressions are rational functions. As a result, algebraic solid models (ASM) are generated by rational functions, a set which is closed under algebraic-Boolean operations.

Figure 1 includes 2-*D* sections of a squaroid and a circle used to demonstrate the algebraic-Boolean operations intersection, union, and difference produced via `implicitplot`. The command `implicitplot` in MapleV2 allows visualization of implicit functions by simply supplying the defining function and the range in each axis. Note that the corners of the squaroid are not very sharp because the defining function is a 4th degree polynomial. Sharper corners can be produced by using a higher degree polynomial to model the squaroid. As mentioned before, the blending effects are most noticeable near the intersections of the objects, giving the resulting sections rounded corners at the intersections.

Another aspect of ASM is the use of rational numbers to specify all coefficients, such as lengths, displacements, and direction cosines [Canny et al. 1992]. All representations and manipulations can then be performed using exact arithmetic. Rational rotation is

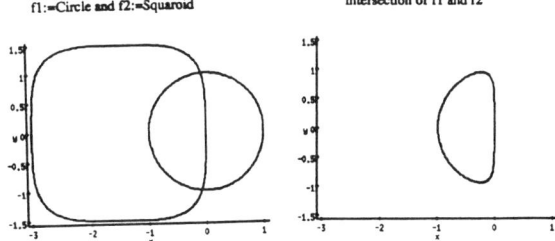

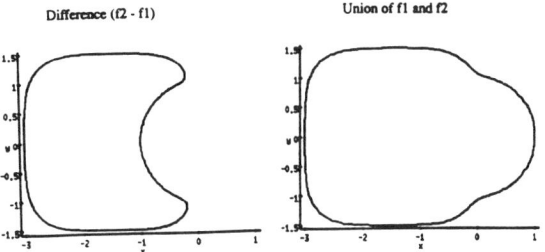

```
> read implicit_solids;
> f1:=Circle;
> f2:=Translate(Squaroid(3/2,4),[-3/2,0]);
> surface1:=Difference([f2,f1],4);
> surface2:=Intersect([f1,f2],4);
> surface3:=Unite([f1,f2],4);
```

FIGURE 1: Boolean Operations on 2-*D* Sections.

demonstrated in Figure 2 and will be discussed in the Coordinate Transformation section.

Primitive Solids

The primitive solids included in `implicit_solids` are sphere, cuboid, and superellipsoid. They can be created by the following commands:

> Sphere;
$$x^2 + y^2 + z^2$$
(A unit sphere centered about the origin)

> Cuboid(4);
$$x^4 + y^4 + z^4$$
(A unit cuboid of degree 4 centered about the origin)

> Superellipsoid(5,6);
$$(x^{1/3} + y^{1/3})^{6/5} + z^{2/5}$$
(A superellipsoid with e1 = 5, e2 = 6)

The solids obtained above can be displayed using Maple's implicitplot3d. Examples of the results of implicitplot3d are shown in Figure 2, 3, and 4. The input arguments to implicitplot3d are very similar to implicitplot; the implicit defining function of the object has to be supplied together with the ranges for each of the axes x, y, and z. Once the object has been displayed, the view of the solid can be rotated by "grabbing" the box that bounds the solid with the mouse and rotating it. The rotated solid can be re-displayed with different options for color schemes, axis locations, and surface characteristics.

Coordinate Transformations

The four basic transformations provided are Translate, Stretch, Magnify, and Rotate. These transformations operate on the defining expressions of the solids. For example, Translate takes a list $[x, y, z]$ as argument where each component of the list represents the corresponding displacement component in that direction. Figure 2 illustrates a unit cuboid of degree 4 stretched, rotated, and then translated. Stretch, similar to Translate, takes a list of the stretch factors in the x, y, and z directions. Note that stretch factors less than 1 produce results equivalent to compressing the solid.

Rotate takes the axis of rotation and the angle of rotation in degrees as arguments. It uses a look-up table, in the form of an array read in from an external file, to find the closest approximation to the transcendental functions in the rotation matrix in terms of rational fraction. The transformation matrix is an exact rotation matrix corresponding to an angle close to the angle of rotation specified. The desired angle is rounded to the nearest degree before a corresponding pair of rational coefficients are found from the look-up table. Higher accuracy is possible using the algorithm by Canny [Canny et al. 1992], but for simplicity only values for integer degrees are generated and stored in the look-up table. By using rational fractions in the computations, the defining functions can be maintained in polynomial form with rational coefficients, and hence algebraic solid models are attained. The result of performing Rotate on the cuboid is shown in Figure 2. Note that the resulting equation after rational rotation contains only rational coefficients and polynomials. The conversion of the desired angle of rotation to rational coefficients is performed in the background, so that the whole operation is invisible to the user. The advantage of this is that the user can concentrate on using the basic ASM commands. The concatenated commands used to generate the extruded cuboid in Figure 2 are listed to demonstrate the compactness of the code.

```
> read implicit_solids;
Warning: new definition for     norm
Warning: new definition for     trace
```

```
> f1:=Cuboid(4);
```
$$f1 := x^4 + y^4 + z^4$$

```
> f2:=Translate(Rotate(Stretch(f1,[1,1,4]),x,10),[3,3,0]);
```
$$f2 := (x-3)^4 + \left(\frac{60}{61}y - \frac{180}{61} - \frac{11}{61}z\right)^4 + \frac{1}{256}\left(\frac{11}{61}y - \frac{33}{61} + \frac{60}{61}z\right)^4$$

```
> implicitplot3d({f1=1,f2=1},x=-3/2..4,y=-3/2..5,z=-5..5,grid=[35,35,35]);
```

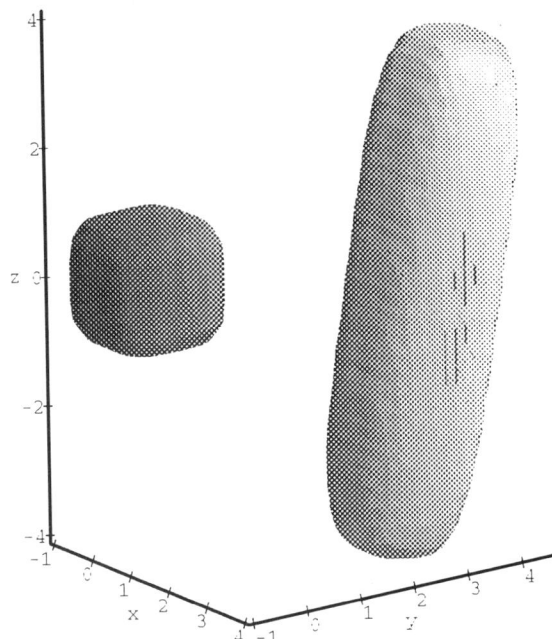

FIGURE 2: Linear Transformations.

Composite Solids

Composite solids can be created by combining primitive solids via algebraic-Boolean operations on them. The three commands Intersect, Unite, and Difference take the defining functions and the blending degree as arguments.

Figure 3 illustrates an example of non-trivial composite solids that can be generated using implicit_solids. The toy truck [Marshall 1986] is created using the basic primitives and the defined Boolean operators. It is displayed using MapleV2's implicitplot3d. The detailed steps of creating and displaying the object are also included in Figure 3. For a complex object such as the truck, the time needed to display the object may be significant if a fine grid is used. The Maple plot shown in Figure 3 requires more than 30 minutes to generate on a HP 700 series workstation. If the object has to be displayed more than once, the plot can be saved to a variable and then re-displayed at a later time. A ray-traced picture of the toy truck is also included in Figure 3 for comparison.

172

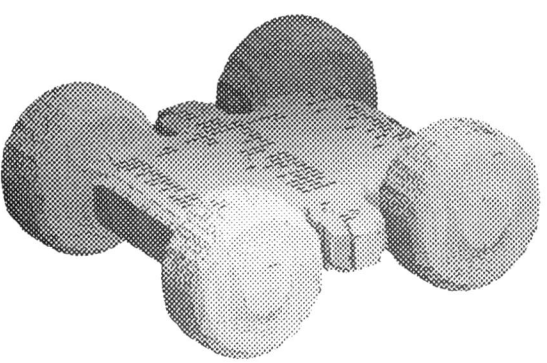

Maple implicit plot.

```
> read implicit_solids;
> wheel:=Extrude(Magnify(Circle,5/8),3/4,6);
> axle:=Extrude(Magnify(Circle,1/8),21/8,6);
> axleholder:=Unite([Stretch(Cuboid(6),[5/4,1/4,7/8]),
        Stretch(Cuboid(6),[1/4,1/4,11/8])]),6);
> truck:=Unite([axleholder,Translate(axle,[1,0,0]),
        Translate(axle,[-1,0,0]),Translate(wheel,[1,0,9/8]),
        Translate(wheel,[1,0,-9/8]),Translate(wheel,[-1,0,9/8]),
        Translate(wheel,[-1,0,-9/8])]),6);
> truck1:=Rotate(truck,x,90);
```

Maple commands.

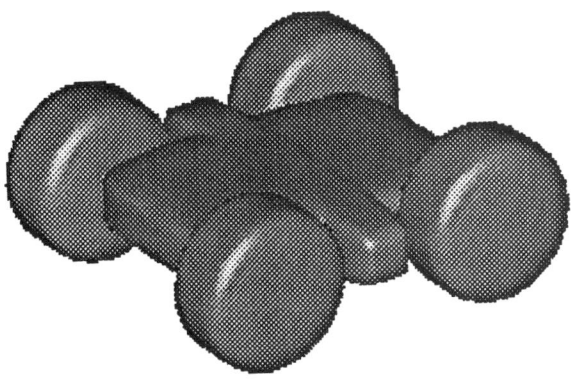

Ray-traced image.

FIGURE 3: An Example of a Composite Solid.

Swept Solids

More complex shapes that do not rely on the basic primitives can also be created by implicit solid modelling. Extrusions, or swept solids, are one such example. These are created by defining a solid and a path along which this solid moves. For simplicity, a one parameter path specified by $(x_0(t),\ y_0(t),\ z_0(t))$ will be considered here.

The family of solids obtained by simultaneously considering the solid associated with all values of t can be thought of as a parametric solid. The surface which lies on and is tangent to a member of this family is called the envelope [Salmon 1934]. This envelope provides the defining function for the desired swept solid.

To develop the envelope, the free parameter is eliminated from the defining function of the original solid and the derivative of that solid's defining function with respect to the free parameter. This can be easily accomplished through the use of MapleV2's `resultant` or Gröbner basis function (`gbasis`).

The power of this method is that the primitive solid does not have to remain constant as it progresses along the path. For example, if the primitive solid is a sphere, the radius of the sphere can be related to the position on the path. Further, a non-symmetrical object, such as an ellipse, can be rotated (using rational rotation techniques) as it is translated.

The envelopes as described allow the free parameter to take on all values in the domain $-\infty < t < \infty$. To create a finite object, the parameter range can be limited by intersection with a range function, R, that satisfies $0 \leq R \leq 1$ on $t_{min} \leq t \leq t_{max}$ and $R > 1$ elsewhere. After intersecting the defining function with the range limiting function, the envelope criterion is applied.

As an example of these methods, consider creating the dumbbell-shaped object displayed in Figure 4 using `implicitplot3d`. The primitive solid was defined as a simple sphere:

$$\text{sphere} = \frac{(x - x_p)^2 + (y - y_p)^2 + (z - z_p)^2}{\text{rad}^2} \qquad (8)$$

In this case, the path was defined to be a straight line along the x axis:

$$x_p = 4t;\ y_p = 0;\ z_p = 0 \qquad (9)$$

The radius of the sphere is determined by a parabolic function:

$$\text{rad} = \frac{7}{4}t^2 + \frac{1}{4} \qquad (10)$$

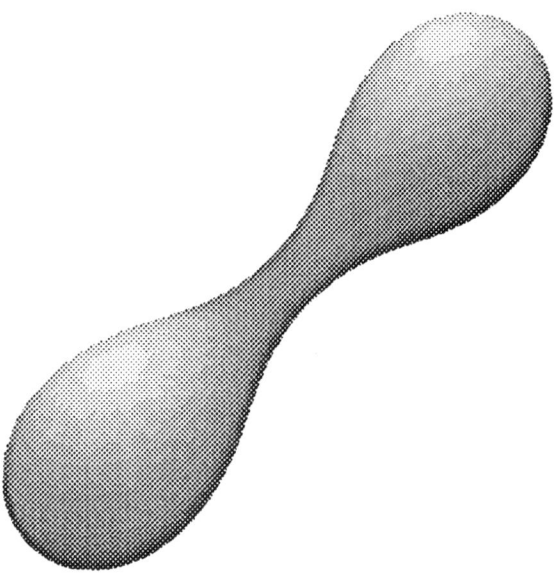

```
> x_path:=4*t;
> y_path:=0;
> z_path:=0;
> rad:=((7/4)*t^2+(1/4));
> sphere:=((x-x_path)^2+(y-y_path)^2+(z-z_path)^2)/(rad^2);
> R:=t^2;
> f:=sphere+R-1;
> nf:=numer(normal(f));
> dnfdt:=diff(nf,t);
> res:=resultant(nf,dnfdt,t);
> readlib(factors);
> res2:=factors(res)[2][1][1];
```

FIGURE 4: An Example of a Swept Solid.

To make the final object finite, a range function was defined as a simple parabola with a value of 1 at $t = \pm 1$ and a minimum value of 0 at $t = 0$:

$$R = t^2 \qquad (11)$$

To intersect this with the sphere's defining function, the standard Boolean intersection is used, but the order is left at one for simplicity.

$$f = \text{sphere} + \text{range} - 1 \qquad (12)$$

A higher order can be used for increased precision at the ends, though at added computational cost. Utilizing MapleV2's `resultant` to remove t from this function, and the function $df/dt = 0$ yields the desired implicit representation of the object.

The swept solid methods just discussed create equations that can be efficiently operated on by implic-

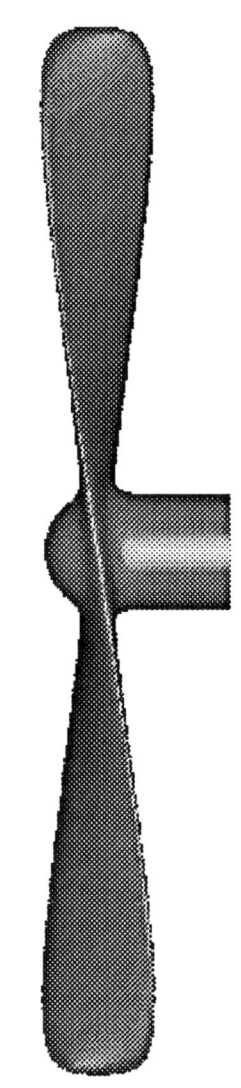

FIGURE 5: Propeller and Shaft.

it_solids. For example, Figure 5 shows a propeller and shaft. The propeller is a swept solid that was generated by rotating an ellipsoid while translating it along a straight line. The shaft was generated by the `Extrude` command in `implicit_solids`. The two were joined together by the `Unite` function.

174

Conclusions

Implicit_solids features compact and easily extensible code working behind a user-friendly interface provided by MapleV2. Users can employ the basic primitives and Boolean set operations provided to create new composite solids and formulate new approximations to Boolean set operations. It is the authors' hope that this package will demonstrate the potential of implicit solid modeling and make ISM available to a broader audience.

Features that the authors are considering adding to the future version of implicit_solids include an option to the rational rotation function which provides an algorithm for computing the rational coefficients for transcendental functions to arbitrary accuracy instead of using a look-up table. This option will produce more accurate angle representation in rational coefficients at the cost of possibly slowing down the function evaluation and plotting processes. The authors find that the union of a computer algebra system and a solid-modeling system is a natural extension of both problem domains.

References

Barr, A.H., "Superquadrics and Angle-Preserving Transformations", *IEEE Computer Graphics and Applications*, Vol. 1, No.1, 1981, pp. 11-23.

Blechschmidt, J.L. and Nagasaru, D., "The Use of Algebraic Functions as a Solid Modeling Alternative: An Investigation", *Advances in Design Automation*, B. Ravani, ed., ASME Design Conference, Chicago, Il., 1990, pp. 33-41.

Canny, J., Donald, B.R., Ressler, E.K., and Rote, G., "A Rational Rotation Method for Robust Geometric Algorithms", *Proc. ACM Symp. on Computational Geometry*, Berlin, 1992.

Char, B.W., Geddes, K.O., Gonnet, G.H., Leong, B.L., Monagan, M.B., and Watt, S.M., *First Leaves: A Tutorial Introduction to MapleV*, Springer-Verlag, 1992.

Foley, J.D.,van Dam, A., Feiner, S.K., and Hughes, J.F., *Computer Graphics: Principles and Practice*, 2nd ed., Addison-Wesley, New York, 1990.

Ganter, M.A. and Storti, D.W., "Object Extent Determination for Algebraic Solid Models", *Advances in Design Automation*, D.L. Hoeltzel, ed., DE-Vol. 44-2, 1992, pp. 275-283.

Ganter, M.A. and Storti, D.W., "Algebraic Solid Modeling: A Renewed Method for Geometric Design", *ASME Resource Book for Innovation in Design Education*, 1992.

Ganter, M.A., Storti, D.W., "Algebraic Methods for Implicit Swept Solids", submitted to *Advances in Design Automation*, 1993.

Marshall, N., *The Great All-American Wooden Toy Book, Rodale Press*, 1986, pp. 126-127.

Requicha, A.A.G. and Voelcker, H.B., "Historical Summary and Contemporary Assessment", *IEEE Computer Graphics and Applications*, Vol.2, No.2, 1982, pp. 9-24.

Ricci, A., "A Constructive Geometry for Computer Graphics", *The Computer Journal*, Vol. 16, No. 2, 1973, pp. 157-160.

Salmon, G., *A Treatise on Higher Plane Curves*, Stechert, NY, 1934, (Photographic Reprint of 3rd Ed. 1879), pp. 73.

E-Mail

ctlim@u.washington.edu (C.T. Lim)
sl1xp@u.washington.edu (M.T. Ensz)
ganter@u.washington.edu (M.A. Ganter)
storti@u.washington.edu (D.W. Storti)

Biographies

C. T. Lim is a Post-Master's student in the mechanical engineering department at the University of Washington. He received a Master's degree in mechanical engineering from the University of Washington in 1992. His primary interests are in the areas of automatic controls, manufacturing, and CAD/CAM.

M. T. Ensz is a Master's degree candidate in the mechanical engineering department at the University of Washington. He received his Bachelor's degree in mechanical engineering from Utah State University in 1992. His primary interests are in the area of CAD/CAM.

M. A. Ganter is an assistant professor of mechanical engineering at the University of Washington. After receiving a Ph.D. from the University of Wisconsin, he has continued to pursue research interests in computational geometry, solid modeling, computer graphics, kinematics, and applications to manufacturing systems automation.

D. W. Storti is an associate professor of mechanical engineering and adjunct professor of applied mathematics at the University of Washington. He received a Ph.D. in theoretical and applied mechanics from Cornell University. His early research in nonlinear dynamics and vibrations led to recent investigations of analytic solid modeling using symbolic computation.

THE ROLE OF A SYMBOLIC PROGRAMMING LANGUAGE IN HARDWARE VERIFICATION: THE CASE OF MAPLE

Farhad Mavaddat
VLSI Group, Dept. of Computer Science, University of Waterloo, Waterloo ON, Canada

ABSTRACT

There is some reluctance towards the use of formal verification methods by the design community. One factor contributing to this lack of enthusiasm is the degree of user sophistication required in representing a design and reasoning about it in most systems, especially those involving theorem provers. To overcome this, we present the use of symbolic programming languages to prove several classes of hardware designs correct.

We start by defining a functional model for the specification of synchronous hardware. Next, we discuss the programming techniques for implementing the model in the symbolic programming language Maple. Given a Maple model (a program) of a design, we execute the program to derive its symbolic behaviour. The Maple system is also used to compare the derived and the reference behaviours. We end by presenting several Maple-based verification examples.

Our contribution to hardware verification is the development of a modeling method within a symbolic programming paradigm, and the ensuing facility for reasoning about certain designs.

1. Introduction

The development of general-purpose symbolic computation systems over the last few years[1-3] suggests their use as a support environment for developing and testing symbolic hardware verification techniques.

Mathematical Computation with Maple V:
Ideas and Applications
Tom Lee, Editor
©1993 Birkhäuser Boston

In[4], we pioneered the use of Maple in proving the correct implementation of hardware whose behaviour is represented as an algorithmic state machine. Here, we continue that discussion by presenting a new methodology and several examples. Our contribution this time, is the ease by which the new methodology helps with representing and reasoning about certain designs.

2. Functional Hardware Model

Gate-level, as well as register-transfer-level, hardware behaviours can be modeled as syntactic extensions of the Lambda calculus. In this section, we present a brief introduction to this modeling technique. Interested readers are referred to [5] for a formal, and in-depth, treatment of this topic, based on denotational semantics.

2.1. Combinational modules

We define an m-input, n-output ($m{\times}n$-put) combinational device D, by

$$D = \lambda\,(\eta_1, \eta_2, \cdots, \eta_m)\,.\,(E_1, E_2, \cdots, E_n). \quad (1)$$

The right side of (1) is a short form for $\lambda\,(\eta_1, \eta_2, \cdots, \eta_m)\,.\,E_i, \quad 1 \le i \le n$, where η_j, $1 \le j \le m$, is the jth input port's value, and $\lambda\,(\eta_1, \eta_2, \cdots, \eta_m).\,E_k, 1 \le k \le n$, is the kth output port's value.

2.2. Sequential modules

The behavior of an $m{\times}n$-put sequential machine B, with q state-variables, is modeled by two combinational circuits defined by

$$B_{cmb}\,(s_1, s_2, \cdots, s_q\,) = \quad (2)$$
$$\lambda\,(\eta_1, \eta_2, \cdots, \eta_m)\,.\,(E_1, E_2, \cdots, E_n)$$

and

$$B_{seq}(s_1, s_2, \cdots, s_q) = \qquad (3)$$

$$\lambda(\eta_1, \eta_2, \cdots, \eta_m) \, . \, (F_1, F_2, \cdots, F_q),$$

where

- $B(s_1, s_2, \cdots, s_q)$ is the behavior of B at state $(s_1, s_2, \cdots, s_q)$ [6].

- $E_1, E_2, \cdots, E_n$ are the n output-port values produced in response to the corresponding input-port and input-state values at all times.

- $F_1, F_2, \cdots, F_q$ are the q next-state values produced in response to the corresponding input-port and input-state values at every step; they are evaluated at the time of transition to the next state.

The two components of B can be combined into a single recursive definition [7]:

$$B(s_1, s_2, \cdots, s_q) = \qquad (4)$$

$$\lambda(\eta_1, \eta_2, \cdots, \eta_m) \, . \, ((E_1, E_2, \cdots, E_n), B(F_1, F_2, \cdots, F_q)).$$

For a more complete discussion of this model, see[8,9].

2.3. Composite modules

An $m \times n-$put composite module $\mathbf{f}^c$ is defined as the interconnection of w sub-modules $\mathbf{f}^1, \mathbf{f}^2, \cdots, \mathbf{f}^w$, and a (hypothetical) $n \times m-$put environment module $\mathbf{f}^0$. The input and output ports of $\mathbf{f}^0$ define the output and input ports of $\mathbf{f}^c$, respectively.

To capture the *net* connections of module

$$\mathbf{f}^j \, (\, m^j \times n^j-put, q^j-state),$$

we write $(Y^j) = \mathbf{f}^j_{cmb}(S^j)(X^j)$ where $Y^j = (y^j_1, y^j_2, \cdots, y^j_{n^j})$ and $X^j = (x^j_1, x^j_2, \cdots, x^j_{m^j})$ are the values of the nets connected to the corresponding ports, and $S^j = (s^j_1, s^j_2, \cdots, s^j_{q^j})$ is the list of state-variables of $\mathbf{f}^j$.

Thus, according to (4), the combined behavior of the module $\mathbf{f}^c$ can be written as

$$\mathbf{f}^c(S^1, S^2, \cdots, S^w) = \qquad (5)$$

$$\lambda(Y^0) \, . \, (\mathbf{rec} \, (\, Y^j = \mathbf{f}^j_{cmb}(S^j)(X^j) \, , \, 1 \le j \le w \,)$$

$$\mathbf{in} \, (\, X^0, \mathbf{f}^c \, (\, \mathbf{f}^j_{seq}(S^j)(X^j) \, , \, 1 \le j \le w \,)) \quad),$$

where **rec** and **in** are defined as in [10].

3. The Maple Model of Hardware

In Sections 2, we outlined a functional formalism for the specification of the behaviour of hardware modules and their composition into larger designs. In this section, we express our formalism in terms of the Maple language constructs [11], and show how a single, hierarchically-defined, Maple function can be used to specify and derive a design's combinational and sequential behaviours.

3.1. The composition model

A typical design, denoted by `j`, composed of sub-modules `f`, `g`, and `h`, is shown in Figure 1.

The general format of a Maple function that represents `j` follows the definition scheme formulated in (5). Of the three sub-modules, `f` and `h` are are sequential with (R1) and (R2,R3) as their respective state-variables. `g` is a combinational module with no state-variables. According to (5), the composite module `j` inherits the sub-modules' state-variables and, therefore, has R1, R2, and R3 as its state variables.

The Maple function corresponding to the design of Figure 1 is shown in Figure 2. The differences between the Maple and the functional models of `j`, as defined in (5), are explained below.

- Line 01 corresponds to $\mathbf{f}^c(S^1, S^2, \cdots, S^w) = \lambda(Y^0)$ in (5). The first parameter, called `mode`, is used to distinguish between the calls made to the combinational (`mode=1`) or the sequential (`mode=0`) behaviours of the device. The state variables (R1,R2,R3) and input variables (W,X) of `j` are on the same side and combined.

- Line 02, required by Maple, and implied by the functional model (5), corresponds to declaring X^j and Y^j, $1 \le j \le w$, in (5). T1,T2, and T3 are temporary variables to be discussed later. The remaining signals, P,Q,R,S,U,Y, and Z, are the internal net names, shown in Figure 1.

- Lines 03 to 12 correspond to **rec** $(Y^j = \mathbf{f}^j_{cmb}(S^j)(X^j)$, $1 \le j \le w)$ in (5). The sub-module connections, in the form of Maple equations, are passed as arguments to the Rec function. A discussion of the Rec function itself is the subject of Section 3.2. The calls g(S,X), f(comb,R1,P,W), and h(comb,R2,R3,Q,R,U) are calls to the Maple models of f, g, and h. Currently, Maple does not support multiple-output functions. Therefore, we use a list to return function values. Each module's

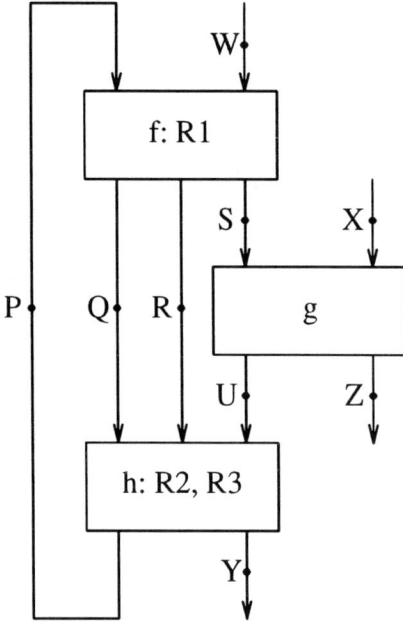

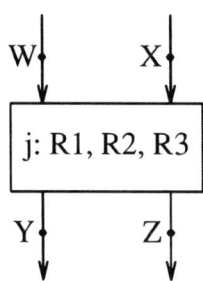

Figure 1: Composition of three submodules f, g, and h (top) into a single composite design j (bottom). The construct 'h:R2,R3' defines module 'h' as having two state variables: 'R2' and 'R3'; similarly for 'f'. Being a combinational element, 'g', has no state-variables.

output list is initially assigned to a temporary variable (T1, T2, and T3 in this example), and then individually assigned to the corresponding net names.

- Line 13, corresponding to in ($X^0, \cdots$) in (5), returns a list of values (in this case [Y,Z]), indicating j's output list.

- Lines 14 and 15 are an iterative implementation of the recursive definition implied by in ($\cdots$, $\mathbf{f}^c$ ($\mathbf{f}^j_{seq}(S^j)(X^j)$, $1 \le j \le w$)) in (5). To update j's state-variables (R1,R2,R3) for the next iteration of the function, sequential calls are made to h and

```
01 j := proc(mode,R1,R2,R3,W,X)
02   local T1,T2,T3,P,Q,R,S,U,Y,Z;
03     Rec(T1 = g(S,X),
04         U = T1[1],
05         Z = T1[2],
06         T2 = f(comb,R1,P,W),
07         Q = T2[1],
08         R = T2[2],
09         S = T2[3],
10         T3 = h(comb,R2,R3,Q,R,U),
11         P = T3[1],
12         Y = T3[2]   )
13     if mode then [Y,Z];
14         else h(sequ,R2,R3,Q,R,S);
15              f(sequ,R1,P,W);
16     fi;
17 end;
```

Figure 2: The Maple function j specifies the behaviour of the composite module 'j' in terms of calls to the behaviours of 'f', 'g', and 'h' submodules.

f. Sequential calls propagate down the design hierarchy, and eventually update the *unit-delay* elements as the only design primitive with a sequential behaviour (Section 3.2).

Note that combinational modules (such as g in example) lack state variables and therefore do not participate in the sequential behaviour of the j module. Furthermore, combinational calls to sequential circuits also do not affect their unit-delay elements.

- Lines 16 and 17, have their usual meanings.

3.2. The underlying system

In[8], we have shown that register-transfer designs can be implemented using only three types of design components, namely: *multiplexors*, *unit-delays*, and *combinational* elements. Although Maple directly supports most of the combinational elements that one might require, such as 'and', 'or', and 'add', we enclose these functions within new functions (typically called 'And', 'Or', or 'Add') that return a list, as required by the Maple model discussed in Section 3.1.

Maple functions Del and Sel, shown in Figure 3, realize the *unit-delay* and *multiplexor* (or *selector*) primitives, respectively. The storage properties of the *unit-delay* primitive are realized through the use of an arrayed global variable, Memo. For the most part, the operations of the Del and Sel functions are clear from their code.

```
Del := proc (mode, N, In)
   if mode then Memo[N]
      else Memo[N] := [ In ] fi;
end;
```
(a)

```
Sel := proc (In1, In2, C)
     [ If ( C, In2, In1) ];
end;
```
(b)

Figure 3: (a) The procedure `Del` realizes the *unit-delay* primitive. A combinational execution of `Del` (`mode=true=1`) simply returns the value in the N-th position of the global array `Memo`. A sequential execution (`mode=false=0`) updates the N-th position by the input to the *unit-delay* element. (b) `Sel` simply calls `If`. Since it is a combinational device, there is no need for any 'mode' control of `Sel`'s execution. Note that the returned values are always lists.

```
Rec := proc() local Rhs,Lhs,Fin,Xi;
   Fin := false;
   while not Fin do
      Fin := true;
      for Xi in args do
         Rhs := eval (op(2,Xi));
         Lhs := op(1,Xi);
         if Rhs <> eval (Lhs, 2)
            then Fin := false fi;
         assign(Lhs, Rhs)
      od;
   od;
end;
```

Figure 4: A possible implementation of `Rec` in Maple. 'op(1, Xi)' and 'op(2,Xi)' refer to the two sides of the equation 'Xi'; 'eval(E,i)' evaluates E recursively to depth 'i'; 'eval(E)' evaluates 'E' to the largest depth possible.

Finally, at the heart of our system is the implementation of the `Rec` function, shown in Figure 4. The `Rec` function receives the design topology as a set of connectivity equations which are referred to as `args`, and which specify the interconnection of the modules according to (5). On invoking `Rec` with a set of connectivity equations, the procedure recursively evaluates and then compares the two sides of every equation; it assigns the right-hand-side expression to the left-hand-side variable until the two sides of every equation evaluate to the same expression. On completion, all left-hand-side variables of a well-defined module, i.e., one with at least one unit-delay element in each connectivity cycle, will be defined only in terms of the input variables, and are returned to the calling environment.

Another useful function at the heart of the system is the `If` definition given in Figure 5. We use this definition rather than the Maple supported 'if' to introduce some sophistication into the evaluation of the *multiplexor (selector)* primitive. `If` tries to recognize the *control* inputs that compare two signals for equality (expressed by the 'Equal' function), and makes a more informed decision about potential equality than simply deciding on the basis of a numerical comparison.

```
If := proc(c, t, e)
   if type(c,function)
      and op(0,c) = Equal
      and type(op(1,c),numeric)
      and type(op(2,c),numeric)
      then if op(1,c) = op(2,c)
         then t else e fi
   elif type(c,function)
      and op(0,c) = Equal
      and op(1,c) = op(2,c) then t
   elif t = e then t
   elif c = true then t
   elif c = false then e
   else 'If(c,t,e)'
   fi
end;
```

Figure 5. A possible implementation of `If` in Maple. `op(0,c)` refers to the function name of the function call `c`. The `type` function works in the obvious way. Expressions enclosed in quotes are returned unevaluated.

4. Interactive Proof Techniques and Examples

To simulate a module and obtain its symbolic behaviour, one has to invoke the Maple package and read the file containing its Maple definition into the Maple environment. After reading the file, one has to initialize each *unit-delay* to its initial symbolic values. Each module may then be executed by entering the corresponding function name with a set of actual symbolic values. To clock a sequential design once, we first invoke the design's model with `mode=1`. This is followed by invoking it with `mode=0`. The effect of sequential behaviours must be derived by inspecting the *unit-delay* elements after executing the corresponding procedures. The behaviour of other modules can be derived in similar ways.

We now outline two proof techniques that can be applied to three categories of digital designs considered in this report. The first technique applies the Maple model to the design of combinational and a special category of sequential circuits, and requires little designer intervention. The second technique, applicable to some other sequential circuits, requires designer intervention and is carried out interactively.

4.1. Proof of combinational circuits

In principle, combinational circuits can be modeled as a straight-line-code of Maple statements ordered by their data dependencies. In practice, the use of the `Rec` function and the design hierarchies simplify the modeling and eliminate the potential for introducing errors in the sequencing of the statements of a large design.

To prove a combinational design correct, the symbolic expression representing the result of evaluating its model is compared with a reference Maple expression using the built-in Maple facilities. The reference expression is either the Maple definition of the expected behaviour (the so-called specification), or obtained by evaluating the Maple model of a design known to be correct. We now present an example of the proof of correctness of a gate-level combinational design.

4.1.1. Boolean combinational proofs

Figure 6 shows a proposed implementation of the 9-input odd function, F, with the following Boolean specification:

$$F(A,B,C,D,E,F,G,H,I) = \qquad (6)$$
$$A \oplus B \oplus C \oplus D \oplus E \oplus F \oplus G \oplus H \oplus I .$$

To derive the implementation's behaviour, its Maple model is called at the Maple command level with suitable symbolic input variables. To prove

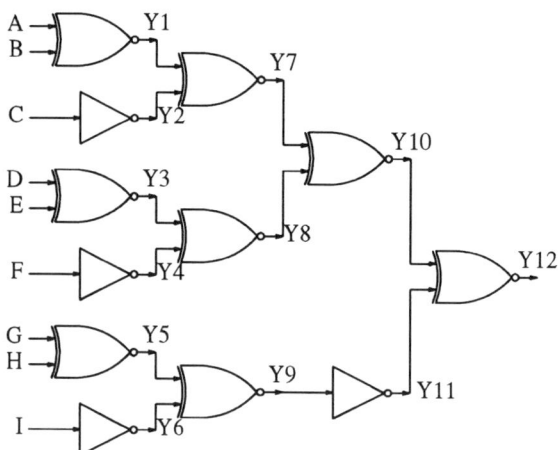

Figure 6: An implementation of the odd function defined by (6). Note that the implementation uses exclusive-nor and inverters instead of the exclusive-ors implied by (6).

the implementation correct, the derived behaviour expression is compared with the specification expression (6) using Maple's `bequal` function. An edited Maple session demonstrating the correctness of the proposed implementation is shown in Figure 6.

4.1.2. Register-transfer combinational proof.

The technique of verifying a combinational register-transfer implementation is similar to the method used in the Boolean domain. The difference is in the use of Maple's `normal` function in comparing the reference to the derived behaviours, as opposed to those of `bequal` or `taut` used in the Boolean domain. We illustrate the techniques with one example.

In Figure 7, two implementations of a register-transfer design are shown. Aside from syntactic differences, the two differ in the modularity of the specifications. To prove the two designs equivalent, their Maple models are evaluated, their resulting polynominal expressions are subtracted from each other, and a single difference-polynominal is formed. If the two designs are identical, the difference polynominal should evaluate to zero. Maple's **normal** function, applied to the difference-polynominal, guarantees a return of zero if the difference-polynominal evaluates to zero[12]. An edited Maple session demonstrating the equivalence of the two implementations is

```
> read OddSpec;
Xor := proc(In1,In2)
      (&not(In1) &and In2) &or
      (In1 &and &not(In2)) end

OddSpec := proc(A,B,C,D,E,F,G,H,I)
    local Y;
    Y = Xor(A,Xor(B,Xor(C,Xor(D,
        Xor(E,Xor(F,Xor(G,I)))))))));
    Y
end

#
> tempSpec:=OddSpec(a,b,c,d,e,f,g,h,i);
#
> read OddImp;
#
Xnor := proc(In1,In2)
      [(&not(In1) &and &not(In2))
      &or (In1 &and In2)] end

Inv := proc(In) [&not(In)] end

OddImp := proc(A,B,C,D,E,F,G,H,I)
    local T1,T2,T3,T4,T5,T6,T7,T8,T9,
        T10,T11,T12,Y1,Y2,Y3,Y4,Y5,
        Y6,Y7,Y8,Y9,Y10,Y11,Y12;
    Rec(T1 = Xnor(A,B),Y1 = T1[1],
        T2 = Inv(C), Y2 = T2[1],
        T7 = Xnor(Y1,Y2), Y7 = T7[1],
        T3 = Xnor(D,E),Y3 = T3[1],
        T4 = Inv(F),Y4 = T4[1],
        T8 = Xnor(Y3,Y4),Y8 = T8[1],
        T5 = Xnor(G,H),Y5 = T5[1],
        T6 = Inv(I), Y6 = T6[1],
        T9 = Xnor(Y5,Y6),Y9 = T9[1],
        T10 = Xnor(Y7,Y8),Y10 = T10[1],
        T11 = Inv(Y9),Y11 = T11[1],
        T12 = Xnor(Y10,Y11),Y12=T12[1]);
    [Y12]
end

> tempImp:=OddImp(a,b,c,d,e,f,g,h,i);
#

> bequal( tempImp[1], tempSpec );
                 true
#
```

Figure 7: The Maple-session aimed at proving the design shown in Figure 5 as a correct implentation of the odd function specified by (6).

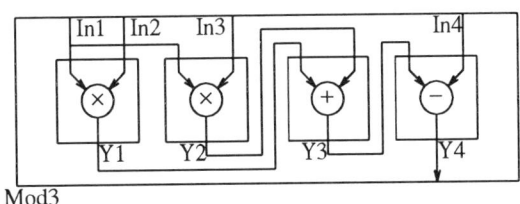

Mod3

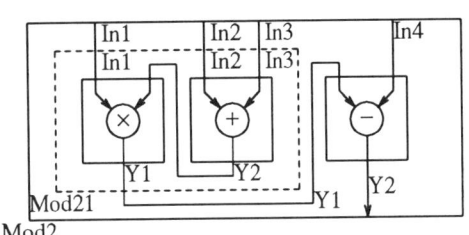

Mod21

Mod2

Figure 8: Two implementations of the (supposedly) same register-transfer behaviour.

shown in Figure 9.

4.2. Proof of the correctness of sequential circuits

Synchronous sequential circuits require one or more clock pulses (steps) to produce their results. We distinguish between those sequential circuits that complete their task after a fixed number of steps, known a priori, and those that require an unknown and possibly variable number of steps before they produce their results.

This distinction helps us to deterministically unroll those with a fixed number of steps and to evaluate them as combinational designs. In Section 4.2.1, while proving the correctness of a design borrowed from[13], we show the ease with which the Maple model can be used in simulating the unrolling step.

However, when the proof procedure is applied to designs with an unknown number of steps, it requires induction and, therefore, the designer's participation. An example of this form of proof (also borrowed from[13]) and the techniques by which the Maple model may help the designer to derive the iteration invariants, is given in Section 4.2.2.

```
#
> read RT_Implementation;
#
Add := proc(In1,In2) [In1+In2] end

Sub := proc(In1,In2) [In1-In2] end

Mul := proc(In1,In2) [In1*In2] end

Mod21 := proc(In1,In2,In3)
    local T1,T2,Y1,Y2;
    Rec(T1 = Mul(In1,Y2),Y1 = T1[1],
        T2 = Add(In2,In3),Y2 = T2[1]);
    [Y1]
end

Mod2 := proc(In1,In2,In3,In4)
    local T1,T2,Y1,Y2;
    Rec(T1 = Mod21(In1,In2,In3),
        Y1 = T1[1],
        T2 = Sub(Y1,In4),
        Y2 = T2[1]);
    [Y2]
end

Mod3 := proc(In1,In2,In3,In4)
    local T1,T2,T3,T4,Y1,Y2,Y3,Y4;
    Rec(T1 = Mul(In1,In2),Y1 = T1[1],
        T2 = Mul(In1,In3),Y2 = T2[1],
        T3 = Add(Y1,Y2),Y3 = T3[1],
        T4 = Sub(Y3,In4),Y4 = T4[1]);
    [Y4]
end

#
> Imp1 := Mod2( a, b, c, d );
#
> Imp2 := Mod3( a, b, c, d );
#
> normal( Imp1[1] - Imp2[1] );
            0
```

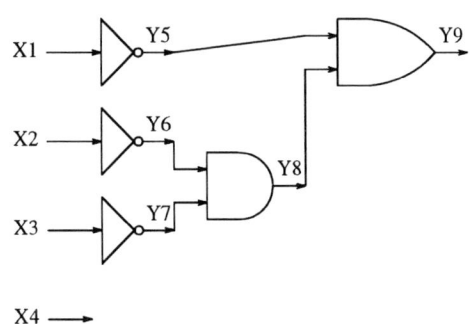

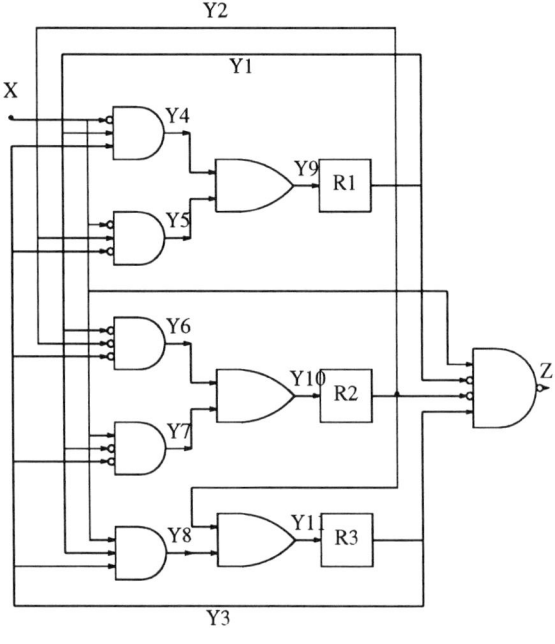

(b)

Figure 9: A Maple session that proves the two designs shown in Figure 7 have the same behaviour.

Figure 10: Two different implementations of the BCD code recognizer. (a) The combinational implementation; inputs are expected in parallel. Note that the X4 input is not used. (b) The sequential implementation; inputs, including X4, are expected in series.

4.2.1. Proof of sequential circuits with a known number of steps

Figures 10.a and 10.b show a combinational and a sequential implementation of a BCD code recognizer, respectively. The combinational design (the reference design) is a straightforward implementation of the BCD code definition. It accepts the code to be recognized in a parallel form. The sequential form is designed to produce the same result in four steps. It accepts the code to be recognized in serial form. We prove that the two behaviours are functionally equivalent.

Both of the designs are expanded to include their input and output sequences. This helps to capture a sequential design's interaction with the environment in a precise form, and minimizes the amount of user intervention between the steps of sequential behaviour. For the sake of uniformity, we also expand combinational designs to include their input and output, necessitating the application of a single pulse to the combinational circuits. To prove the two designs functionally equivalent, we evaluate their respective Maple models and use Maple's `bequal` function to test their identity.

To evaluate each design, the input unit-delays are initialized, the circuit's Maple model is evaluated, and the output unit-delays are inspected. The symbolic expression assigned to the output unit-delay is the circuit's symbolic output.

To evaluate the sequential design, the circuit's Maple model has to be stepped four times, passing the state values between the state transitions via Maple's global variables. A lightly edited Maple session demonstrating the equivalence of the two implementation is shown in Appendix I.

4.2.2. Sequential circuits with an unknown number of steps

In this section, we consider designs whose behaviour can be modeled by a single while-loop. Although this may seem restrictive, it can be shown to be an appropriate model for a large number of application-specific designs.

What distinguishes these designs from those requiring a fixed, a priori known, number of steps is the need for a status output (called **cond**) that signals the step at which the data output(s) (called **result**) produce their results. In effect, we have to show

$$\textbf{cond}(\text{ k }) \supset \textbf{result}(\text{ k }) \equiv \textbf{specification}, \quad (7)$$

where **cond**(k) and **result**(k) are the symbolic output expressions at **cond** and **result** at the k step. As the value of k for which (7) holds is not known, this leads to an unpredictable number of computation steps. J. L. Paillet has shown that by induction on the number of steps we should be able to reason about such behaviour[13], but the derivation of the induction steps is left to the designer.

We now discuss the use of the Maple model of the design as an aid to deriving the induction step. To this end, starting from an initial configuration, we clock the model through a few steps, comparing the *unit-delay* (symbolic) values (i.e., the state variables) at successive steps. Based on the succession of the observed values, we (try to) conjecture the behaviour values at the k-th and the $(k+1)$-th steps. This is followed by assigning the conjectured expressions, for the k-th step, to corresponding *unit-delay* element, in effect, bringing the system into the k-th step.

Finally, we use the design's Maple model to verify the conjectures. This is done by stepping the updated design (the one with *unit-delays* representing the k-th step) by one step, to the $(k+1)$-th step, and comparing the new *unit-delay* values with those of $(k+1)$-th step, also conjectured and stored in the temporary variables. Should the conjectures be correct, the anticipated $(k+1)$-th step expressions and those derived by stepping the model should be equivalent. In addition to the `normal`, `bequal`, and `taut` facilities discussed before, the Maple environment provides additional useful facilities for the management of induction steps, including that of `subs` which helps with unit-delay updates to the k-th step, thereby minimizing the potential of mistakes in the handling of large expressions.

Figure 11 shows a sequential design, proposed by Gordon[14], for computing the product of two integers held at its inputs X and Y throughout the computation. The output 'DONE' is initially a '1', and the device is stepped until the 'DONE' output becomes '1' again, at which time the product of input values is available at 'OUT'. A Maple session demonstrating the correctness of the design, using the method discussed above, is shown in Appendix II. During the session we show that **cond**(k) $\equiv (a{-}k = 0) \vee (b = 0)$, and **result**(k) $\equiv (\text{if}(a = 0), 0, b) \times k$. Using the values of **cond** and **result** at the kth step, one can easily prove (7), shown by:

$$(a{-}k = 0) \vee (b = 0) \supset (\text{if}(a = 0), 0, b) \times k$$

$$\equiv a \times b. \quad (8)$$

183

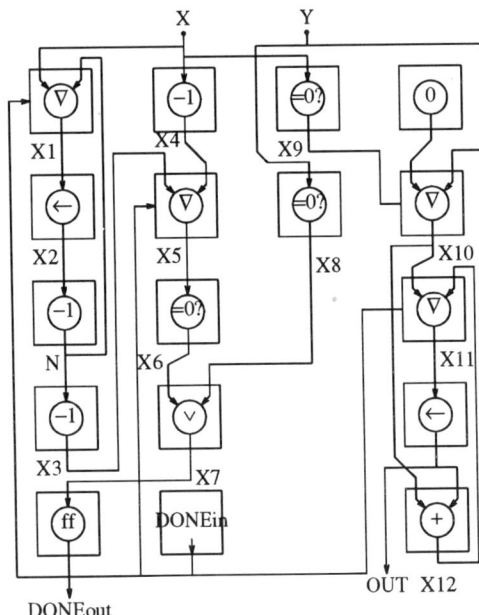

X

Y

X1

X4

X9

X2

X5

X8

X10

N

X6

X11

X3

X7

DONEin

OUT X12

DONEout

Figure 11: A proposed implementation of the multiplier. Boxes with ∇ and ← inside, represent the *multiplexor* and *unit-delay* elements, respectively. Functional units are shown by boxes with their common algbraic symbol inside. Boxes with no inputs, represent constants. A box with string 'ff' inside, represents a flipflop.

5. Conclusions

Tautology checkers derive their power from their restriction to boolean theory, and therefore can exploit the specialization of that theory. As long as we can specify our design within that theory, a tautology checker is going to give us a very effective way to determine correctness. On a negative side, we have to cast every aspect of the design that concerns us into a binary representation. There is a trade-off between a more fully automatic proof capability and the degree to which the engineer must interpret the representation over which the proof is done.

We followed a similar procedure in the Maple demonstrations, casting our designs in the arithmetic theory for which Maple is specialized. One reason this worked well was that the aspect of design with which we were concerned was readily represented by an arithmetic formula.

Mechanized logics, such as the Boyer-Moore[15] and HOL[16] systems, are designed to construct theories and reason about them. They allow the user to specify data types; they provide a way to describe how elementary values are interpreted without saying how those values are represented.

We believe that the formal-methods community should retreat from the notion that the design domain can be encompassed by a single logical framework. The reason is that the proof aids, because they are so general, are too weak to be practical. Efforts to compensate by making theorem provers 'programmable' aren't going to pay off soon enough.

We also believe that too few people realize that the research in the area is as much about methodology as it is about proving circuits. The methodology is intimately related to the modeling framework and the reasoning tools: identifying this relationship presents a primary research goal.

6. Acknowledgements

The author would like to acknowledge the support by the Natural Sciences and Engineering Research Council of Canada under research grant OGP0005515, and the University of Waterloo funding and computing facilities used to carry out some of the work reported here. The author also gratefully acknowledges K. Geddes, B. Leong, M. Monagan, and A. Pathria of Waterloo's Maple group, for their assistance with Maple related issues at different stages of this work.

7. References

1. Fitch, J., "Solving Algebraic Problems with REDUCE," *Journal of Symbolic Computation*, **1** pp. 211-217 (1985).

2. Pavelle, R. and Wang, P. S., "MACSYMA from F to G," *Journal of Symbolic Computation*, **1** pp. 69-100 (1985).

3. B.W. Char, G.J. Fee, K.O. Geddes, G.H. Gonnet, and M.B. Monagan, "A Tutorial Introduction to Maple," *Journal of Symbolic Computation*, **2** pp. 179-200 (1986).

4. Mavaddat, F., "Inductive Assertions on Algorithmic State Machines A Maple-Based Register-Transfer-Level Proof System," pp. 257-266 in *Formal VLSI Correctness Verification, VLSI Design Methods-II*, ed. L.J.M. Claesen, North Holland, Amsterdam, Netherlands (1990).

5. Gordon M., "A Model of Register Transfer Systems with Application to Microcode and VLSI Correctness," CSR-82-81, University of Edinburgh, Dept. of Computer Science, Edinburgh, Scotland (March 1981- revised May 1982).

6. Gordon, M. J. C., "The Denotational Semantics of Sequential Machines," *Information Processing Letters*, **10**(1) pp. 1-3 (February 1980).

7. Milner, R., "Processes: A Mathematical Model of Computing Agents," pp. 157-173 in *Logic Colloquium '73*, ed. H.E Rose and J. C. Shepherdson, North Holland Publishing Company, Amsterdam, Holland (1975).

8. Mavaddat, F., "Designing and Modeling VLSI Systems at Register Transfer Level," *International Journal of Computer Aided VLSI Design*, **2** pp. 281-314 (1990).

9. Mavaddat, F., "A Functional Model of Register-Transfer Designs," *Proceedings of Twentieth Annual Pittsburgh Conference on Modeling and Simulation*, (May 1989).

10. Landin, P. J., "The Mechanical Evaluation of Expressions," *Computer Journal*, **6**(4) pp. 308-320 (Jan. 1984).

11. Char, B. W., Geddes, K. O., Gonnet, G. H., Monagan, M. B., and Watt, S. M., *MAPLE First Leaves: A Tutorial Introduction to Maple*, Springer-Verlag, New York (1992).

12. Geddes, K.O., Gzapor, C., and Labahn, G., *Algorithms for Computer Algebra*, Kluwer Academic Publisher, Boston (1992).

13. Paillet, J. L., "A Functional Model for Description and Specification of Digital Devices," pp. 21-42 in *From HDL Descriptions to Guaranteed Correct Circuit Designs*, ed. D. Borrione, Elsevier Science Publishers B. V. (North Holland), Amsterdam, netherlands (1987).

14. Gordon, M. J. C., "Why Higher Order Logic is a Good Formalism for Specifying and Verifying Hardware," Technical Report No. 77, Computer Laboratory, University of Cambridge, Cambridge, England (September 1985).

15. Boyer, R. S. and Moore, J. S., *A Computational Logic*, Academic Press, New York (1979).

16. Gordon, M. J. C., "HOL A Machine Oriented Formulation of Higher Order Logic," Technical Report No. 68, Computer Laboratory, University of Cambridge, Cambridge, UK (1985).

8. Biography

Farhad Mavaddat received the PhD degree from Imperial College of Science and Technology, London, England. From 1968 to 1979 he was with Arya Mehr University of Technology, Tehran, Iran. Since 1979 he has been with Department of Computer Science, University of waterloo, Waterloo, Ontario, Canada. His present research interests include the use of formal methods in design. He can be reached by electronic mail sent to 'fmavaddat@math.uwaterloo.ca'.

```
> with(logic);                              Memo[1] := [false]
> comb := true;
> sequ := false;                  > Memo[2] := [false] ;
#                                            Memo[2] := [false]
#  In the interest of brevity
#  definitions of logicr-gates of  > Memo[3] := [false] ;
#  all types are removed from this          Memo[3] := [false]
#  and the next page.
                                   #
Bcd_comb := proc(mode, X1, X2, X3, X4)  #
  local T1,T2,T3,T4,T5,T6,T7,T8,T9,  # Next the design is pulsed for
  Y1,Y2,Y3,Y4,Y5,Y6,Y7,Y8,Y9;     # four complete cycles.
  Rec(                             #
  T5 = Not1(X1), Y5 = T5[1],       > Bcd_sequ(comb, 1, 2, 3, X4);
  T6 = Not1(X2), Y6 = T6[1],                   [true]
  T7 = Not1(X3), Y7 = T7[1],
  T8 = And11(Y6, Y7), Y8 = T8[1],  > Bcd_sequ(sequ, 1, 2, 3, X4);
  T9 = Or11(Y5, Y8), Y9 = T9[1]);             [false]
  if mode then [Y9]
    else [];                       #
  fi;                              > Bcd_sequ(comb, 1, 2, 3, X3);
end;                                          [true]
#
> Exp1 := Bcd_comb(comb,X1,X2,X3,X4);  > Bcd_sequ(sequ, 1, 2, 3, X3);
    Exp1 := [not (X1 and (X2 or X3))]           [true]

> F1 := convert( Exp1[1], toinert);  #
    F1 := &not (X1 &and (X3 &or X2))  > Bcd_sequ(comb, 1, 2, 3, X2);
#                                             [true]
Bcd_sequ := proc(mode, R1, R2, R3, X)
  local T1,T2,T3,T4,T5,T6,T7,T8,T9,  > Bcd_sequ(sequ, 1, 2, 3, X2);
  T10,T11,T12,Y1,Y2,Y3,Y4,Y5,Y6,        [X3 or X2 and not X3]
  Y7,Y8,Y9,Y10,Y11,Z;
  Rec(                             #
  T4 = And011(X,Y1,Y3), Y4 = T4[1],  > Exp2 := Bcd_sequ(comb,1,2,3,X1);
  T5 = And010(X,Y2,Y3), Y5 = T5[1],  Exp2 := [not (X1 and (X2 or X3)
  T6 = And000(Y1,Y2,Y3), Y6 = T6[1],     and (X3 or X2 and not X3))]
  T7 = And100(X,Y1,Y3), Y7 = T7[1],
  T8 = And111(X,Y1,Y3), Y8 = T8[1],  > Bcd_sequ(sequ, 1, 2, 3, X1);
  T9 = Or11(Y4,Y5), Y9 = T9[1],         [X1 and not (X2 or X3) and
  T10 = Or11(Y6,Y7), Y10 = T10[1],      (X3 or X2 and not X3)]
  T11 = Or11(Y2,Y8), Y11 = T11[1],
  T12 = Nand1001(X,Y1,Y2,Y3),       > F2 := convert( Exp2[1], toinert);
  Z = T12[1],                       F2 := &not (X1 &and ((X3 &or X2) &and
  T1 = Del(comb,R1,Y9),  Y1=T1[1],      (X3 &or (X2 &and (&not X3)))))
  T2 = Del(comb,R2,Y10), Y2=T2[1],
  T3 = Del(comb,R3,Y11), Y3=T3[1]);  #
  if mode then [Z]                  # Next the two boolean expressions
    else Del(sequ, R1, Y9);         # assigned to F1 and F2 are compared
         Del(sequ, R2, Y10);        # for equality.
         Del(sequ, R3, Y11);        #
  fi;                               > bequal(F1, F2);
end;                                             true
#
> Memo[1] := [false] ;             > quit;
```

Appendix II

```
> with(logic);

ztest := proc(In)
        if type(eval(In),numeric)
        then [evalb(eval(In) = 0)]
        else '[In = 0]' fi
    end

Multip := proc(mode,R1,R2,F1,X,Y,DONE)
  local T1,T2,TN,T3,TDONE,T4,T5,T6,
  T7,T8,T9,T10,T11,TOUT,T12,X1,X2,X3,
  X4,X5,X6,X7,X8,X9,X10,X11,X12,N,OUT;
  Rec(T1='Sel(N,X,DONE)',X1 = T1[1],
      T2='Del(comb,R1,X1)',X2 = T2[1],
      TN='Dec(X2)',N = TN[1],
      T3='Dec(N)',X3 = T3[1],
   TDONE='Del(comb,F1,X7)',
      T4='Dec(X)',X4 = T4[1],
      T5='Sel(X3,X4,DONE)',X5 = T5[1],
      T6='ztest(X5)',X6 = T6[1],
      T7='Or11(X6,X8)',X7 = 'T7[1]',
      T8='ztest(Y)',X8 = T8[1],
      T9='ztest(X)',X9 = T9[1],
      T10='Sel(Y,0,X9)',X10 = T10[1],
      T11='Sel(X12,X10,DONE)',
      X11=T11[1],
     TOUT='Del(comb,R2,X11)',
      OUT=TOUT[1],
      T12='Add(X10,OUT)',
      X12=T12[1]);
  if mode then [OUT,TDONE[1]]
      else Del(sequ,R1,X1);
      Del(sequ,R2,X11); Del(sequ,F1,X7)
  fi
end
#
> comb := true; sequ := false;
> Memo[1] := [a]; Memo[2] := [0];
> Memo[3] := [true];
#

> Multip(sequ, 1, 2, 3, a, b, true );

> Memo[1]; Memo[2]; Memo[3];
              [a]
          [If(a = 0, 0, b)]
        [Or(a - 1 = 0, b = 0)]


> Multip(sequ, 1, 2, 3, a, b, false);

> Memo[1]; Memo[2]; Memo[3];
             [a - 1]
         [2 If(a = 0, 0, b)]
        [Or(a - 2 = 0, b = 0)]
```

```
> Multip(sequ, 1, 2, 3, a, b, false);

> Memo[1]; Memo[2]; Memo[3];
             [a - 2]
         [3 If(a = 0, 0, b)]
        [Or(a - 3 = 0, b = 0)]

#  Conjecturing and inputting values.
#
> Memo[1] := subs(-2=-(k-1),Memo[1]);
       Memo[1] := [a - k + 1]

> Memo1    := subs( k = k+1, Memo[1]);
       Memo1 := [a - k]

> Memo[2] := subs(  3 = k, Memo[2]);
     Memo[2] := [If(a = 0, 0, b) k]

> Memo2    := subs( k = k+1, Memo[2]);
     Memo2 := [If(a = 0, 0, b) (k + 1)]

> Memo[3] := subs( -3 = -k, Memo[3]);
     Memo[3] := [Or(a - k = 0, b = 0)]

> Memo3    := subs( k = k+1, Memo[3]);
     Memo3 := [Or(a - k - 1 = 0, b = 0)]

> Multip(sequ, 1, 2, 3, a, b, false);
     [Or(a - k - 1 = 0, b = 0)]
#
#    Tesing the conjecutres.
#
> Memo[1]; Memo1;
             [a - k]
             [a - k]

#
> Memo[2]; Memo2;
     [If(a=0,0,b) + If(a=0,0,b) k]

     [If(a = 0, 0, b) (k + 1)]

> normal(Memo[2][1] - Memo2[1]);
             0
#
> Memo[3]; Memo3;
     [Or(a - k - 1 = 0, b = 0)]
     [Or(a - k - 1 = 0, b = 0)]

#
#    Therefore, the conjectured
#    expressions are correct and
#    as a result (8) holds.
```

A SYMBOLIC CSG SYSTEM WRITTEN IN MAPLE V

Darren Thompson, Tom Trias, Laurence Leff
Department of Computer Science, Western Illinois University, Macomb IL, USA

We have designed a working 2-D Constructive Solid Geometry (CSG) system in MAPLEV, which is currently comprised of more than 1600 lines of code. Our research in this area is important for several reasons. To begin with, it is the first user expandable CSG system. We designed the system to allow the user to add their own definitions of curves and primitives. It is also the first symbolic CSG system. The advantage of this is that the user can change the initial values of the parameters to obtain different sizes of the same conceptual object. An inequality list, generated by our program, will ensure that the object is conceptually the same. This is important as to enable shape optimization of objects. The generated inequality list will also eliminate needless computations as parameters are varied by simply checking whether or not the new input values satisfy the inequality list. Finally, this method of symbolic evaluation serves as a paradigm for other areas of science and engineering.

Mathematical Computation with Maple V:
Ideas and Applications
Tom Lee, Editor
©1993 Birkhäuser Boston

I. Explanation of CSG

In CSG, objects are represented as the applications of the boolean operations **difference**, **intersection**, and **union** to regular sets. Some examples of regular sets are squares and limaçons in two dimensions or spheres and cylinders in three dimensions. The user performing CSG enters information about objects using primitives such as squares, limaçons, etc. The user then indicates how they are to be combined into the new object by taking the **difference**, **intersection**, or **union** of these primitives. A more detailed explanation of CSG can be found in [TIL80], [TIL84], and [TRIAS93].

II. The CSG Boundary Representation and M Algorithms

We were able to draw from Robert Tilove's work in [TIL77] and [TIL81] to help us design our algorithms to convert our CSG expressions to a boundary representation. Table I shows the CSG Boundary Representation algorithm. If given a CSG expression of the form

```
CSG_EXP:=rectangle(a,b,side1,
        side2) &intersect
        rectangle(c,d,side3,
        side4)
```

```
function edgelist (T);
 /*T is a CSG tree
    (returns a list of
    edges on the boundary
    of the object
    represented)*/
  edgelist <- nil
  for each primitive P_i
   for each edge E of P_i
    (EinS, EonS, EoutS)
     <- M[E,T]
    merge (EinS, EonS,
    EoutS) with edgelist
   end /*for*/
  end /*for*/
  remove edges in or out
  of object leaving edges
  on
end /*edgelist*/
```

Table I   CSG Boundary
              Representation
              Algorithm [TIL81]

our code will take this
expression, containing these
two rectangles, and place them
in a CSG tree. The first
rectangle will be the left-hand
son, the second rectangle will
be the right-hand son, and the
intersect operator will be the
parent of the two. This tree
will then be passed to the
function **edgelist**. **Edgelist**
will then break each primitive
into segments and pass each
segment, along with the tree
passed to **edgelist** to the
function **M**. Table II shows the
M algorithm. This algorithm
takes a segment and classifies
it against every primitive
within the CSG tree. **M** then
computes whether the segment is

in, **on**, or **off** the primitive.
Figure 1 demonstrates how the
segment is bubbled up through
the tree. We perform a table
look-up in order to obtain the
new values produced when
performing one of the
difference, **intersection**, or
union operations of a segment
and a primitive. The segment
being classified against each
primitive is rectangle **A**'s
bottom segment.

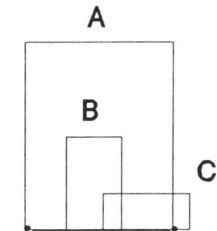

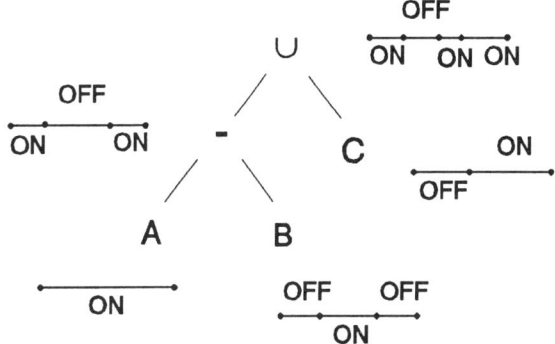

Figure 1 Classifying a segment
            against primitives

    Below is a partial listing
of our procedure **combine1** to
perform the **difference** task.

                    .
                    .

```
1  if operation = `&difference`
2     then
3       if status1 = `on` and
4          status2 = `on` then
5        if bit1 = bit2 then
```

189

```
    function M (Edge,T);

    /*T is a CSG Tree

    this procedure returns an edge divided into segments with
    on, off, and in
    for those segments with the on indication, there is also a
    bit that indicates whether the region is above or below the
    segment*/

    if T is a primitive then
      perform a primitive classification
    else
      Result_L = M (left hand son of T);
      Result_R = M (right hand son of T);
      Produce a merged result for this Edge
        using Result_L and Result_R
```

Table II   CSG Edge Merge Procedure [TIL80]

```
6            RETURN(['off',nil])        35 fi;
7         else                          36 fi;
8           RETURN(['on',bit1]);
9        fi;                                      .
10       elif status1 = 'on' and                  .
11             status2 = 'in' then
12          RETURN(['off',nil]);
13       elif status1 = 'on' and
14             status2 = 'off' then
15          RETURN(['on',bit1]);
16    elif status1 = 'in' and
17          status2 = 'on' then
18          RETURN(['on',
19            revbit(bit2)]);
20    elif status1 = 'in' and
21          status2 = 'off' then
22          RETURN(['in',nil]);
23    elif status1 = 'in' and
24          status2 = 'in' then
25          RETURN(['off',nil]);
26    elif status1 = 'off' and
27          status2 = 'off' then
28          RETURN(['off',nil]);
29    elif status1 = 'off' and
30          status2 = 'in' then
31          RETURN(['off',nil]);
32 elif status1 = 'off' and
33          status2 = 'on' then
34          RETURN(['off',nil]);
```

Notice that in Figure 1, there is a **difference** operation between rectangles **A** and **B**. The segment being classified is completely on **A** and only partly on **B**. Our program compares these two classifications and bubbles up the result. When our program sees **on** and **off**, it looks at lines 13-15 and returns the result **on**. When it compares **on** and **on** and sees that both **bits 1** and **2** are the same, it returns **off**. Both **bits** are the same since the rest of rectangle **A** is above the segment being classified and the rest of rectangle **B** is also above the segment being classified. The same procedure takes place for every other segment in each primitive. The result is then unioned together to form the resulting object.

190

III. The Classification of Segments

We classify general segments against one another by a procedure called **segment_classify**. This procedure is important to our program for two reasons. First, it generalizes the routine to classify two segments against one another. Secondly, this routine has the power of MAPLEV. Our main code is designed to handle simple cases which have been preprogrammed for efficiency (e.g., classifying a horizontal segment against a vertical or vice versa). On the other hand, **Segment_classify** is designed to handle any curve types which can be defined in MAPLEV. In order to classify two segments correctly, all points where they intersect need to be found. If MAPLEV can find those points of intersection, then **segment_classify** has the ability to return the correct values.

Segment_classify takes two arguments: segment to be classified (**segment1**) and segment to be classified against (**segment2**). This procedure will return two important lists of lists. The first is an inequality list which will provide the necessary constraints for the segments. It also enables the user to change the values of the parameters and MAPLEV makes sure the constraints hold. If the constraints do not hold, the user will be made aware that their input will produce an object conceptually different from the object they initially entered. For example, a plate with a hole in it may be divided into two pieces as the value of the parameter giving the dimension of the hole becomes large enough. Some systems supporting parameters in mechanical engineering CAD do not generate the needed constraints. Thus, physically unrealizable shapes may be produced as parameters vary--an example is given in [GOS88]. These constraints correspond to the "side conditions" of [BOT82]. The second list of lists produced by **segment_classify** is the **return_value_list**. **Return_value_list** contains the following information:

```
[[item1 - atom stating whether
         the segment1 is on,
         off, or in segment2,
  item2 - if item1 is on, this
         will be 0 or 1 (as
         per cbit) otherwise
         null,
  item3 - [left-hand-point -
         symbolic value,
         grounded value],
  item4 - [right-hand-point -
         symbolic value,
         grounded value]]
```

We use the **gsolve** routine from the Grobner Basis package to solve for the intersections. We did this in order to find all intersections when dealing with polynomials.

IV. Database.z

Database.z is a file which contains all definitions of

curves and primitives. We designed our system to be user expandable. We knew we couldn't come up with all possible curve and primitive definitions to meet every user's needs; therefore, the user has the ability to define their own curve or primitive and add it to this file. All the user has to do is put their definition in the same format as our definitions. Below is a definition of a quadratic curve.

```
#------------------------------
#FT'S OF TYPE5
#QUADRATIC CURVES

ft[5,vars] :=[a,b,c];
ft[5,vs] := 3;
ft[5,v1] := a;
ft[5,v2] := b;
ft[5,v3] := c;
ft[5,peqx] := p;
ft[5,peqy] := a*p^2+b*p+c;
```

We assigned each curve an **ft** type. This curve definition just happened to be fifth in the list, so the number has no other special meaning. **Vars** gives the variables required to define an infinite curve of this type. We found that defining all curves parametrically allows greater generality of programming. A rectangular primitive definition is shown below.

```
#===== RECTANGLE ======
prim[rectangle,arg]:=xll,yll,
                     s1,s2;

prim[rectangle,1,ft]:=2;
# => VERTICAL FT
prim[rectangle,1]:=xll,yll,
                   yll+s2;
```

```
prim[rectangle,2,ft]:=1;
# => HORIZONTAL FT
prim[rectangle,2]:=yll+s2,
                   xll,xll+s1;

prim[rectangle,3,ft]:=2;
# => VERTICAL FT
prim[rectangle,3]:=xll+s1,
                   yll+s2,yll;

prim[rectangle,4,ft]:=1;
# => HORIZONTAL FT
prim[rectangle,4]:=yll,xll+s1,
                   xll;
```

The **xll**, **yll**, **s1**, and **s2** variables are assigned the lower-left x- and y-coordinates, length, and width. With these values, MAPLEV can construct the rectangle.

V. Advantages of Programming in MAPLEV

The key advantage of MAPLEV in our research was the ability to represent arbitrary expressions and curves. This would allow the engineer to use the full mathematical repertoire of MAPLEV to represent the curves in the object they are designing, rather than being limited to a specific set as in a conventional CAD system. We also rely on the ability of MAPLEV to solve two equations simultaneously to determine the points of intersection of these segments.

Another advantage is our ability to generate a plot with a single statement. It would take us countless hours to program this ourselves. We need the ability to plot the

```
                                 •
                                 •
                                 •
1    paralistgen:=proc(segm,segplace)
2    local temp,eqlist;
3     eqlist:=[];
4     for temp from 1 to nops(ft[segplace[1],vars]) do
5         eqlist:=[op(eqlist),ft[segplace[1],vars][temp] =
                     segplace[temp+1]];
6     od;
7      paralist:={op(paralist),subs(eqlist,[ft[segplace[1],
                 peqx],ft[segplace[1], peqy],
                 p=segm[3][2]..segm[4][2]])};
8     end; # paralistgen
                                 •
                                 •
9    map(proc(n) if assigned(op(2,n)[2]) then
                    map(paralistgen,op(2,n)[2],op(1,n))   fi;
         end,op(op(segmentsets))):
10   h:=plot(paralist):
11   for i from 4 to nops(h) do
12       h:=subs(op(2,op(i,h))=BLACK,h):
13   od:
14   print(h);
                                 •
                                 •
```

Text Box 1   The code used to generate plots

boundary of the object after the CSG algorithm has been run. **Paralistgen**, which does this for us, is shown in Text Box 1.

The **map** function, in line 9, processes each element of **segmentsets**. The second element of each table entry represents the segments that are on the object. Thus, **op(2,n)** is the list of segments that require plotting for this segment set. **Op(1,n)** represents the entry in the hash list. This is a vector consisting of a **ft-type** and the grounded values for each variable in the definition of that curve.

Lines 4 and 5 of **paralistgen** will replace each variable in the **ft-type** with the corresponding value for this segment. **Ft[segplace[1], vars]** will provide the list of variables, other than **x** and **y**, contained within the definition of a curve. For example, the parabola defined above will have **a**, **b**, and **c** for this. When this loop is finished, **eqlist** will contain a list of substitutions that will convert an equation of this type to the specific curve on which these segments lie. **Segplace[temp+1]** will represent the specific value for the **temp**'th variable in the curve definition. Thus, when **temp** is 2, the code would substitute a specific number for **b** in the curve. In line 7, the substitutions defined by **eqlist** are applied to the parametric equation for the segment. **Ft[segplace[1],peqx]** is the equation for **x** in the parametric representation of the curve--and similarly for **ft[segplace[1],peqy]**.

Another example of the ability of Maple to easily define powerful data structures is given in Text Box 2. It shows an entry in the **segmentsets** data structure, the most widely-used data structure in our system. It shows one of the eight entries generated when the CSG expression defined in section VII is given to this program. The first 1 in **[1,1]** specifies the **ft** type of the segment (e.g., horizontal, vertical).

```
[[1, 1]= table([
  1 = [[c, 1],
       [a + side1, 2],
       [c + side2, 3]]
  2 = [[on, 1,
       [a + side1, 2],
       [c + side2, 3]]]
  3 = [8]])
```

**Text Box 2   Part of
               segmentset's
               data structure**

A segment of **ft-type** 1 is a horizontal line. The second 1 gives the grounded value to define an infinite curve of this type. Therefore, this segment is a horizontal line with a **y-value** of 1. **op(1,n)** is used to index **[1,1]**. The

next important set of information used to plot the object is the second entry of the table. This segment is **on** the object. This list is in the format of the **return_value_list** described above. There are eight entries in the **segmentsets** data structure since the object generated has eight sides--each of which is a line segment of a unique infinite line.

MAPLEV also saves us a considerable amount of time when writing code. This is because MAPLEV contains some of the LISP functions (e.g., map, subs, member). For example, we frequently had to substitute the nominal values of the parameters into various structures and expressions. Section VII shows example input data for a symbolic CSG situation. Observe the definition of the **bs** vector which describes the nominal values for the parameters of the actual CSG expression. We use the procedure **subsbs**, listed below, to substitute these nominal values for the corresponding parameters for us.

```
subsbs := proc(expr);
subs(op(op(bs)),expr);
end;
```

If we had to write the code to perform these operations, it would take us a significant amount of time.

In many instances of our program, we had to keep what computer scientists refer to as a dictionary. For example, we

194

had to keep track of all the segments that potentially overlapped. More specifically, we had to keep and process together all of the segments of the same infinite curve.

Another example of this is keeping track of the information for each primitive. We simply write **prim[rectangle,2,ft]** to obtain the **ft-type** for the second segment of the rectangle. Note that the first parameter of the **prim hash_list** gives the primitive name, which of course can change as the user adds a new primitive. The second gives the number of the segment within the primitive (there may be an arbitrary number of segments within each primitive). The third parameter tells the system we need the information on the **ft-type**.

All this clearly could have been done using linked lists or binary trees in a conventional programming language such as C or PASCAL. However, both of these subfunctions and many other similar subfunctions in our program, would have had to be implemented separately, each requiring dozens of lines in a more typical language.

VI. Disadvantages to
     Programming in MAPLEV

As mentioned in the previous section, we were relying on the MAPLEV **solve** function to intersect the curves composing the primitive shapes. Unfortunately, it has

let us down on what would seem to us as simple cases. We represent all curves as parametric equations. Below, is an example of our difficulties with the important job of intersecting a circle or an arc with a line. **Solve** and **fsolve** are unable to find all the roots to trigonometric functions.

```
f:=cos(p)=p1;
                f := cos(p) = p1

g:=sin(p)=0;
                g := sin(p) = 0

solve({f,g},{p,p1});
                {p = 0,  p1 = 1}

fsolve({f,g},{p,p1},
  {p=Pi/2..3*Pi/2,p1=-1.5..0});
                {p = 0,  p1 = 1.}

cos(Pi);
                -1
sin(Pi);
                0
```

Notice that a range was placed on **p** and **p1** to help **fsolve** find the solution {**p = Pi, p1 = -1**}. Unfortunately, this has slowed down our progress in dealing with trigonometric curves. We are now taking an alternative approach by using polynomials since MAPLEV is more successful in finding their roots.

We have wasted a considerable amount of time debugging our code due to the lack of strong typing in MAPLEV. For example, we have an array of lists called **segments**. Many of our subroutines required a segment as one of the parameters. In some cases, the segment itself would be passed. In other cases, the subscript to the **segments** array would be passed. Needless to say, when the subscript was passed where the segment itself was needed, or vice versa, difficult-to-find bugs were encountered. This problem would be analogous to passing the pointer to a structure instead of a structure to a subroutine. A good typed system catches any errors.

MAPLEV does not have a stand-alone compiler. This has been a serious drawback since our program is 1600 lines and growing. Every time we want to make changes in our code, we have to call **fred(<filename>)**, make changes and then exit. When we exit, MAPLEV has to read in every line of code. This can take some time if the user is on a slow machine. One way to reduce the time it takes to read in a file is to break it up into sub-files. The user then has to keep track of when they need to read in a file to re-initialize variables. If the user is in the middle of a **fred** session, and wants to test the syntax before coding it in the program, they have to exit **fred**. This means MAPLEV will have to read in the entire file. The user can then test their idea; if it works, they then have to go back into **fred** to edit their file. Therefore, a stand-alone compiler would enhance the MAPLEV programming environment tremendously.

MAPLEV is not programmer-friendly when it comes to errors. When there is an error in our 1600-line program, it takes us more time than necessary to find it due to the lack of good error checking and/or a better way of informing the programmer that an error has occurred. When **F3** is pressed to exit **fred**, MAPLEV then reads in the file. The entire file scrolls up the screen; this can happen too quickly to watch on a fast computer. If the word "error" written with an arrow pointing to the error is not seen, an attempt is made to run the program and the mistake in syntax is not found until incorrect results are produced. If the syntax is wrong, MAPLEV will ignore that statement, if not the whole procedure. The programmer has the alternative to send the output to a file, but then they would have to inspect the output file and check to see if there are any syntax errors.

We do know that there is an add-on MAPLE-callable package for the C programming environment; unfortunately, we have not had the opportunity to try it. This package would provide us a command-line compiler, tighter error checking, and allow us to define data types.

VII. Results

Shown below are the results of a simple example input and output from our code **unioning** two squares. The input to our system was as follows:

```
csg:=square(a,b,side1) &union
    square(c,d,side2);
bs[a]:=0;
bs[b]:=0;
bs[side1]:=2;
bs[c]:=1;
bs[d]:=1;
bs[side2]:=2;
```

This input defines **square1** as a square with its lower-left **x**- and **y-coordinate** at **(0,0)** and its side being length 2. **Square2** has **x-** and **y-coordinates (1,1)** and its side is also length 2. The output from our system was the inequality list shown below.

```
[b, a, a + side1, b, a,
 a + side1, a + side1,
 a < a + side1, b + side1, a,
 a + side1, b + side1,
 b + side1, a, a + side1,
 a + side1, a < c,
 c < a + side1, c, c, d, d,
 d + side2, c, d, d + side2,
 d < b + side1,
 b + side1 < d + side2,
```

```
 c + side2, d, d, d + side2,
 c + side2, d, d + side2,
 d < d + side2, a + side1, b,
 b + side1, a + side1,
 a + side1, b, b + side1,
 b + side1, b < d,
 d < b + side1, a, b,
 b + side1, a, b, b + side1,
 b + side1, b < b + side1, d,
 d, c, c, c + side2, d, c,
 c + side2, c < a + side1,
 a + side1 < c + side2,
 d + side2, c, c,
 c + side2, d + side2, c,
 c + side2, c < c + side2]
```

Keep in mind that the entries **b**, **a**, etc. imply **b > 0**, **a > 0**, etc. You may notice some redundancy in the list; we have not yet dealt with the task of deleting redundant inequalities. A method of doing this, and arguments regarding time complexity can be found in [LEFF90]. A plot of the resulting object is shown below.

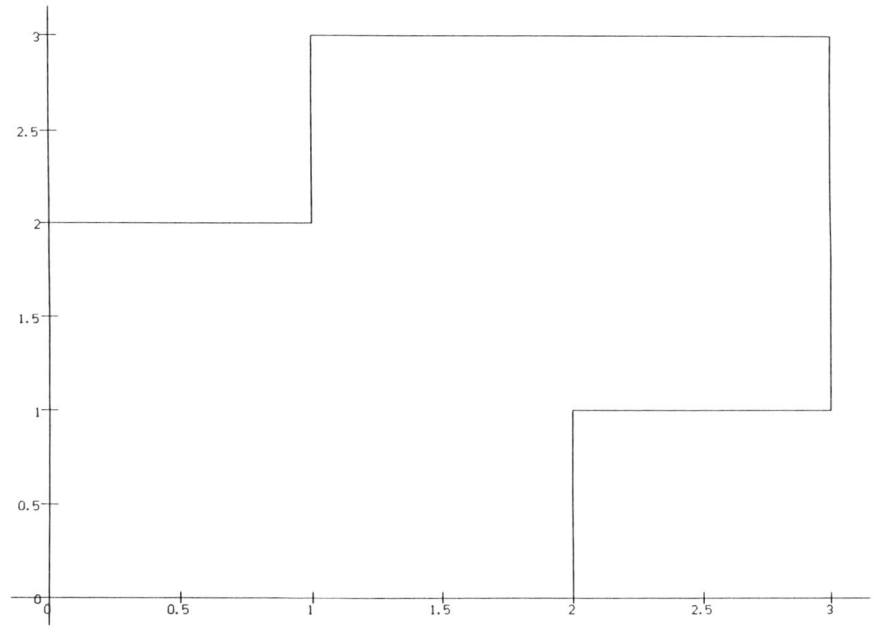

References

BOT82   Botkin, M.E., Shape Optimization of Plate and Shell Structures, *AIAA Journal*, Vol. 20, pages 268-273, FEB 1982.

GOS88   Gossard, D.C., Zuffante, R.P., and Sakurai, H., Representing Dimensions, Tolerances and Features in MCAE Systems, *IEEE Computer Graphics and Applications*, Vol. 8, No. 2, pages 51-59, MAR 1988.

LEFF90   Leff, L., *Symbolic Finite Element Analysis and Constructive Solid Geometry*, Ph.D. Thesis Southern Methodist University, 1990.

TIL77   Tilove, R.B., *A Study of Geometric Set-Membership Classification*, TM-30, Production Automation Project, University of Rochester, Rochester, New York, NOV 1977.

TIL80   Tilove, R.B., Exploiting Spatial and Structural Locality in Geometric Intersection Problems, *IEEE Transactions on Computers*, Vol. C-29, No. 10, pages 874-883, OCT 1980.

TIL81   Tilove, R.B., *Exploiting Spatial and Structural Locality in Geometric Modelling*, TM-38, Production Automation Project, University of Rochester, Rochester, New York, OCT 1981.

TIL84   Tilove, R.B., A Null-Object Detection Algorithm for Constructive Solid Geometry, *Communications of the ACM*, Vol. 27, No. 7, pages 684-694, JUL 1984.

TRIAS93   Trias, T.S., Kyaw, M., Thompson, D., Leff, L., and Malik, Z., Towards a Symbolic Math CAD/CAM System, submitted to American Society of Mechanical Engineers 1993 Joint Conference on Engineering Systems Design and Analysis.

About the Authors

Darren Thompson is currently a senior at Western Illinois University and is planning to attend graduate school to earn his Ph. D. in Computer Science. He has presented papers in this area of research at The Illinois State Academy of Science, 1993 and TIMS/ORSA, 1993.

Tom Trias did the work while an undergraduate at Western Illinois University. He anticipates graduating and proceeding to a graduate program shortly.

Laurence L. Leff, Ph.D. is an Assistant Professor of Computer Science and is active in applying symbolic mathematics techniques to Computer Aided Design and Finite Element Analysis for mechanical engineers.

Please send all correspondence to: Western Illinois University % Darren Thompson, 447 Stipes Hall, Macomb, IL 61455.